MECHANICAL & ELECTRONIC

DRAFTING FOR TRADES & INDUSTRY

JOHN A. NELSON

DRAFTING FOR TRADES & INDUSTRY

MECHANICAL & ELECTRONIC

JOHN A. NELSON

DELMAR PUBLISHERS
COPYRIGHT © 1979
BY LITTON EDUCATIONAL PUBLISHING, INC.

10 9 8 7 6 5 4 3 2

LIBRARY OF CONGRESS CATALOG CARD NUMBER: 77-91450
ISBN 0-8273-1846-4

Printed in the United States of America
Published simultaneously in Canada by
Delmar Publishers, A Division of
Van Nostrand Reinhold, Ltd.

DELMAR PUBLISHERS • ALBANY, NEW YORK 12205
A DIVISION OF LITTON EDUCATIONAL PUBLISHING, INC.

PREFACE

Each manufactured object in the world around us began as a simple idea, a mental image which later was set down on paper in the form of working drawings. The engineer, architect, or designer furnished the major calculations, general specifications, and preliminary sketches of the object. The drafter combined these elements into a finished drawing showing all of the details. It is this conversion of a mental image to a finished drawing that makes up the world of drafting.

Drafting for Trades and Industry — Mechanical and Electronic is designed to develop the student's technical skills in mechanical and electronic drawing. The student will learn to design, calculate, and dimension detail, assembly, and schematic drawings. The text explains how to compute limits, tolerances, and fits. It also demonstrates how to draw such features as welds, fasteners, springs, cams, and electronic components.

While traditional texts present new material in narrative form, *Drafting for Trades and Industry — Mechanical and Electronic* involves the student with extensive hands-on experience that applies drafting theories and develops skills. However, this is not simply a workbook. Each topic is developed through a progression of practice exercises that are augmented with explanations and suggestions to help the student overcome areas of difficulty.

In addition to the practice exercises, each unit includes a pretest and unit review. The pretest allows both the student and the instructor to determine where special attention is required. The unit review gives the instructor a means of evaluating the student's performance. Answers to all practice exercises are included within the units. Answers to pretests and unit reviews appear in a separate Instructor's Guide.

By providing the answers to practice exercises in the student's edition, this drafting program may be used in the traditional classroom or in an independent study program. Students using this book may advance at their own speed. The teacher is present to give guidance, explanations, and demonstrations when needed by the student. In this way, each individual can practice drafting techniques until they are mastered. The combination of instructor, proper instructional materials, and practice is the ideal way to develop a skill.

Drafting for Trades and Industry — Mechanical and Electronic is designed for the student who has mastered basic drawing techniques. *Drafting for Trades and Industry — Basic Skills* provides this experience. Once basic drawing techniques are mastered, the student is ready to specialize in a particular field of drafting.

There are four areas on concentration in the *Drafting for Trades and Industry* series:

- *Drafting for Trades and Industry — Architectural*
- *Drafting for Trades and Industry — Civil*
- *Drafting for Trades and Industry — Technical Illustration*
- *Drafting for Trades and Industry — Mechanical and Electronic*

For a comprehensive overview of drafting, the student should complete all the books in the series. However, the student may be ready to seek employment as a mechanical or electronic drafter after completing *Basic Skills* and the material in the *Mechanical and Electronic* concentration.

ABOUT THE AUTHOR

John A. Nelson is an experienced drafting instructor. He has spent 11 years working in design and drafting departments in industry and has 14 years experience as a drafting instructor. He has a degree from the College of San Mateo in California and has done post-graduate work at Foothill College in California and the University of Tennessee.

CONTENTS

INTRODUCTION

Drafting for Trades and Industry — Mechanical and Electronic covers technical skills in mechanical and electronic drafting. Each topic is developed through a series of practice exercises which implement the information presented in the unit. There are no lectures. All material is illustrated and explained in the unit. The teacher provides additional guidance, explanations, and demonstrations when needed. In this way, students are free to work at their own pace and practice the skills until they are mastered.

Each student is expected to advance as rapidly as possible. The student should be familiar with the drafting reference books available in the classroom and library. These books provide alternate methods of drawing and additional exercises for the student who is having difficulty in a particular area of drafting. Be sure each point is fully understood. If something is unclear, ask the instructor.

Unit Instructions

The *objective* explains the purpose of the unit. It outlines what information and skills the student will learn as a result of completing the exercises contained in the unit.

Most units begin with a *pretest* covering the material contained in the unit. Those who pass the pretest can go directly to the next unit. Those who are unfamiliar with the material or fail the pretest move through the unit and complete all the practice exercises.

Related terms are designed to teach the language of drafting. They are completed in the spaces provided after the information is read. Related terms should be thought of as study notes.

Each unit contains *information* which must be read before actual practice in drafting skills can take place. In some cases, information contains background material, while others contain instructions on how to perform particular skills or solve specific problems.

Information is followed by *practice exercises*. After completing each exercise, the work should be compared to the answer provided in the unit *before proceeding* to the next exercise or topic. Any questions should be discussed with the instructor before going on.

The *unit review* consists of an examination covering everything contained in the unit. It is taken after all work in the unit is completed. Read the instructions carefully, as many reviews have time limits within which the work must be done. If the unit review is successfully completed within the time limit, the student proceeds to the next unit of study.

In this book, inch measurements are followed by comparable metric values rounded off in millimetres. This metric value is not an exact equivalent of the inch measurement but a conversion to a realistic metric size. Where precision is required, metric values are stated as precisely as necessary.

Though inch and metric values are included on most drawings in this text, this does not mean that dual dimensioning is practical in every situation. Dual dimensioning gives the instructor the option of introducing the student to the metric system of measurement. The instructor should specify which measurement system to use in the text. The metric values appear in parentheses. For example, "1 inch (25)" means that the comparable metric value of 1 inch is 25 millimetres.

The *flow chart* provides a graphic illustration of how the student progresses through the unit. By following the arrows, one can quickly see the procedure to follow to successfully complete each unit.

At the end of each unit review, there is room to record two things. First, the instructor may initial approval for the work completed in the unit. Second, the student may indicate that the unit's completion was recorded on the progress chart.

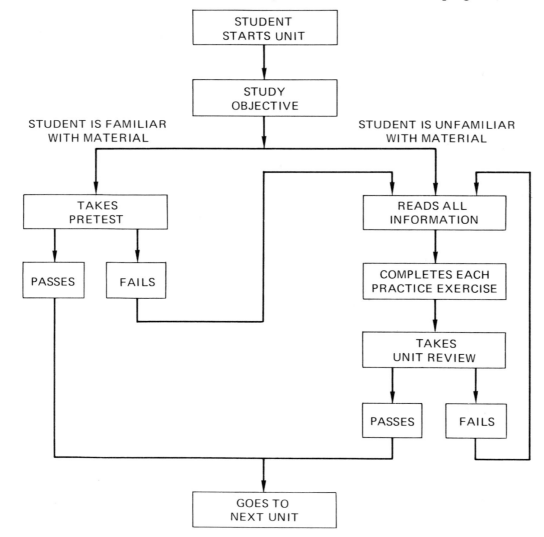

Flow Chart

PROGRESS CHART

The *progress chart* is a graph used to record the progress made by the student through the course. By using this chart, students know exactly where they are in the program and what units remain in order to complete all the work in the time allotted.

The numbers at the bottom of the graph indicate the number of school days. The unit titles are listed at the left. The "recommended progress" line indicates the average length of time the student should spend on each unit.

To record a unit's completion, place a dot where the unit title and the day the unit is completed meet. For example, *Unit 5 — Springs* is completed on the 20th day on the recommended progress line.

If the dot falls above the recommended progress line, then the student is ahead of schedule. If the dot falls below the line, the student is running behind schedule and should adjust the rate of progress if all the units are to be completed in time.

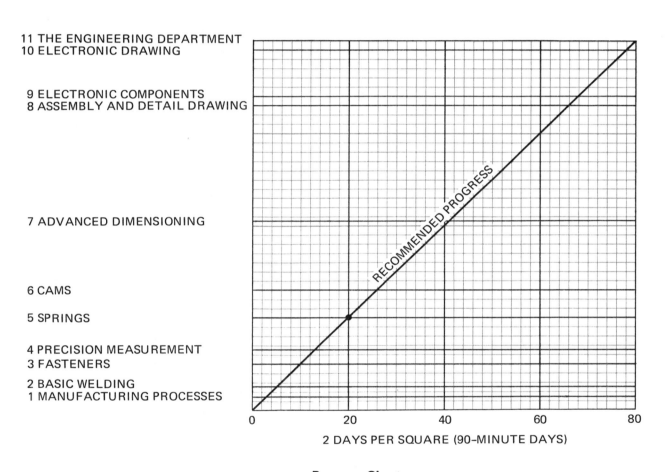

Progress Chart

APPLICATION CHART – DRAFTING FOR TRADES AND INDUSTRY

Occupational Title

UNIT NO.	UNIT TITLE	Tracer/Detailer	Architectural Drafter	Cartographer	Electronic Drafter	Technical Illustrator	Mechanical Drafter	Electro-Mechanical Drafter	Tool Designer	Drafting Teacher
Drafting for Trades and Industry – BASIC SKILLS										
1	Equipment	▮	▮	▮	▮	▮	▮	▮	▮	▮
2	Lettering	▮	▮	▮	▮	▮	▮	▮	▮	▮
3	Drawing Techniques	▮	▮	▮	▮	▮	▮	▮	▮	▮
4	Geometric Construction	▮	▮	▮	▮	▮	▮	▮	▮	▮
5	Multiview Drawings	▮	▮	▮	▮	▮	▮	▮	▮	▮
6	Basic Isometrics	▮	▮	▮	▮	▮	▮	▮	▮	▮
7	Section Views	▮	▮	▮	▮	▮	▮	▮	▮	▮
8	Descriptive Geometry	▮	▮	▮	▮	▮	▮	▮	▮	▮
9	Auxiliary Views	▮	▮	▮	▮	▮	▮	▮	▮	▮
10	Developments	▮	▮	▮	▮	▮	▮	▮	▮	▮
11	Basic Dimensioning	▮	▮	▮	▮	▮	▮	▮	▮	▮
12	Careers in Drafting	▮	▮	▮	▮	▮	▮	▮	▮	▮
Drafting for Trades and Industry – ARCHITECTURAL										
1	House Considerations		▮							▮
2	House Construction		▮							▮
3	Windows, Doors, Fireplaces, and Fixtures		▮							▮
4	Stair Layout		▮							▮
5	Structural Members and Loading		▮							▮
6	Working Drawings		▮							▮
Drafting for Trades and Industry – CIVIL										
1	Maps and Surveys		▮	▮						▮
2	Plotting		▮	▮						▮
3	Surveying		▮	▮						▮
4	Contour Maps		▮	▮						▮
Drafting for Trades and Industry – TECHNICAL ILLUSTRATION										
1	Mechanical Lettering		▮			▮				▮
2	Advanced Isometrics		▮			▮				▮
3	Perspective Drawing		▮			▮				▮
4	Airbrush Techniques		▮			▮				▮
Drafting for Trades and Industry – MECHANICAL AND ELECTRONIC										
1	Manufacturing Processes						▮	▮	▮	▮
2	Basic Welding						▮	▮	▮	▮
3	Fasteners						▮	▮	▮	▮
4	Precision Measurement						▮	▮	▮	▮
5	Springs						▮	▮	▮	▮
6	Cams						▮	▮	▮	▮
7	Advanced Dimensioning						▮	▮	▮	▮
8	Assembly and Detail Drawing						▮	▮	▮	▮
9	Electronic Components				▮			▮		▮
10	Electronic Drawing				▮			▮		▮
11	The Engineering Department				▮		▮	▮		▮

▮ Required

The chart suggests the minimum number of units to cover for a particular field of interest. However, the final course content is the decision of the individual instructor.

MANUFACTURING
PROCESSES

OBJECTIVE

The student will learn about the basic materials used in industry, machine tool operations, and the casting process in order to know how various parts are made.

PRETEST

45-minute time limit

1. List four kinds of steel.

2. Why is cast iron so popular in the manufacturing of machine parts?

3. What kind of steel is SAE 3010 and what is it usually used for?

4. Name two major kinds of case hardening tests. Which is read directly from a scale on the tester?

5. What is the difference between strength and plasticity?

6. List three important drawing notations which must be placed on a drawing if the part contains a metal.

7. Which kind of grinder is used to grind:
   a. a flat surface?

   b. a round part?

   c. a hole?

8. What is a turret lathe?

9. List the five basic machine tool operations.

10. Explain what a draft is in the casting process.

11. List three functions of rounds and fillets.

12. What is the difference between a rib and a web?

RELATED TERMS

Give a brief definition of each term as progress is made through the unit.

Alloy _____

Metallurgy _____

Strength _____

Plasticity _____

Ductility _____

Malleability _____

Elasticity _____

Brittleness _____

Toughness _____

Fatigue limit _____

Aging _____

Quenching _____

Tempering _____

Annealing _____

Case hardening _____

Extrusion _____

Drilling _____

Turning _____

Planing _____

Grinding _____

Milling _____

Lathe _____

Turret _____

Broach _____

Casting _____

Split pattern _____

Core support _____

Molding board _____

Drag _____

Cope _____

Flask _____

Sprue _____

Riser _____

Shrinkage allowance _____

Draft _____

Bosses and pads _____

Lugs _____

METALLURGY

A drafter must know about the behavior, characteristics, and properties of metals. This unit gives a very general working knowledge of metals, but much more on-the-job study must be done by the drafter.

Pure metals by themselves are usually too soft and weak to be used for machine parts. Thus, alloys are used. An *alloy* is simply a mixture of metals and chemical elements.

Materials must be carefully chosen to give the best working life of the part to be made and still be in line cost-wise with competition. In industry, most companies have one or more metallurgists who works with the engineering department to assist in the selection of correct metal or alloy for the design and function of each machine part. *Metallurgy* is the art and science of separating metals from their ores and preparing them for use.

CHARACTERISTICS OF METALS AND ALLOYS

The composition of metal and various chemical elements regulates the mechanical, chemical, and electrical properties of that metal. The following terms describe certain characteristics and capabilities associated with metals and alloys:

Strength is the ability to resist deformation.

Plasticity is the ability to withstand deformation without breaking. Usually hardened metals have strength but are very low in plasticity. They are brittle.

Ductility describes how well a material can be drawn out. This is an especially important characteristic for wire drawing and metal shape forming.

Malleability is the ability of a metal to be shaped by hammering or rolling.

Elasticity is the ability of metal to be stretched and then return to its original size.

Brittleness is a characteristic of metal to break with little deformation.

Toughness describes a metal that has high strength and malleability.

Fatigue limit is the stress, measured in pounds per square inch, at which a metal will break after a certain number of repeated applications of a load has been applied.

Conductivity describes how well a metal transmits electricity or heat.

Corrosion resistance describes how well a metal resists rust. Note that *rust* adds weight, reduces strength, and ruins the overall appearance of a metal.

HEAT TREATMENT

Heat treatment of metals and alloys provides certain desirable properties. Listed are a few basic terms associated with heating treatment:

Aging is a process that takes place slowly at room temperature.

Quenching is the process of cooling a hot metal in water or oil. In special cases it can be quenched in sand, lime, or asbestos rather than liquid in order to slow down the cooling process.

Tempering or *"drawing"* is the process of reheating and cooling by air. Tempering increases toughness, decreases hardness, relieves stress, and removes some brittleness.

Annealing is the process of softening a hardened metal so it can be shaped or machined. Annealing also removes internal stresses which cause warping and other deformations.

Case hardening hardens only the outer layer of the material. In this process, the outer layer of the metal absorbs carbon or nitrogen, thus hardening the outer layer.

Hardness testing is done by two methods: the Brinell hardness test or the Rockwell hardness test. The *Brinell method* uses a hardened steel ball that is pressed into the surface of the metal under a given pressure or load. The depth the ball goes into the metal surface is measured by a microscope and is converted into a hardness reading. The *Rockwell method* is very much the same except the hardness value is read directly from a scale attached to the tester.

DRAWING NOTATIONS

It is important that any part to be hardened is noted on the drawing. This notation must include the material, the heat-treating process, and the hardness test method and number.

TYPES OF METALS

There are two classifications of metal: (1) *ferrous*, or those that contain iron; and (2) *nonferrous*, or those that do not contain iron.

Ferrous Metals

Cast iron is widely used for machine parts. It is relatively inexpensive and easily cast into most any shape. It is a hard metal, strong, and has a good wearability. It responds very easily to almost all heat treating processes but tends to be brittle. *Malleable iron* is used where parts are subject to shock.

Steel is an alloy composed of iron and other chemical elements. It is important to remember that carbon content in steel regulates the properties of various types of steel. There are four classes of steel: carbon, alloy, stainless, and tool. A standard system of designating steel has been established by the American Iron and Steel Institute (A.I.S.I.) and the Society of Automotive Engineers (S.A.E.) to describe the type of steel to be used. The drafter must indicate the A.I.S.I.-S.A.E. numbers as indicated in figure 1-1. SAE1010 steel, for example indicates carbon steel with approximately 0.10% carbon.

PROPERTIES, GRADE NUMBERS & USAGES			
Class of Steel	*Grade Number	Properties	Uses
Carbon - Mild 0.3% carbon	10xx	Tough - Less Strength	Rivets - Hooks - Chains - Shafts - Pressed Steel Products
Carbon - Medium 0.3% to 0.6% carbon	10xx	Tough & Strong	Gears - Shafts - Studs - Various Machine Parts
Carbon - Hard 1.6% to 1.7%	10xx	Less Tough - Much Harder	Drills - Knives - Saws
Nickel	20xx	Tough & Strong	Axles - Connecting Rods - Crank Shafts
Nickel Chromium	30xx	Tough & Strong	Rings Gears - Shafts - Piston Pins - Bolts - Studs - Screws
Molybdenum	40xx	Very Strong	Forgings - Shafts - Gears - Cams
Chromium	50xx	Hard W/Strength & Toughness	Ball Bearings - Roller Bearing - Springs - Gears - Shafts
Chromium Vanadium	60xx	Hard & Strong	Shafts - Axles -Gears - Dies - Punches - Drills
Chromium Nickel Stainless	60xx	Rust Resistance	Food Containers - Medical/Dental Surgical Instruments
Silicon - Manganese	90xx	Springiness	Large Springs

*The first two numbers indicate type of steel, the last two numbers indicate the approx. average carbon content — 1010 steel indicates, carbon steel w/approx. 0.10% carbon.

Fig. 1-1 Properties, grade numbers, and usage of steel alloys

Nonferrous Metals

Copper is soft, tough, and ductile. It is a good conductor of both electricity and heat.

Brass is an alloy of copper (copper/zinc) and very workable, tough, and ductile.

Bronze is another alloy of copper (copper/tin). It is a serviceable, strong, and tough metal.

Aluminum is very malleable, ductile, and a good conductor of electricity and heat. It is very light in weight. Aluminum cannot be heat treated; thus, to increase its hardness, other alloys and elements must be added.

Magnesium is perhaps the lightest metal used today. It is a good conductor of electricity and heat, nonmagnetic, easily machined, but highly inflammable while machining.

SHAPES OF METALS

Metals are purchased from the manufacturer in standard shapes and sizes. The drafter must use the correct callouts when indicating what material is to be used, figure 1-2.

Extrusion is a method of forming very odd or special shapes, similar to squeezing toothpaste from a tube. The round opening is like a die (the required shape) and the toothpaste represents the metal to be shaped.

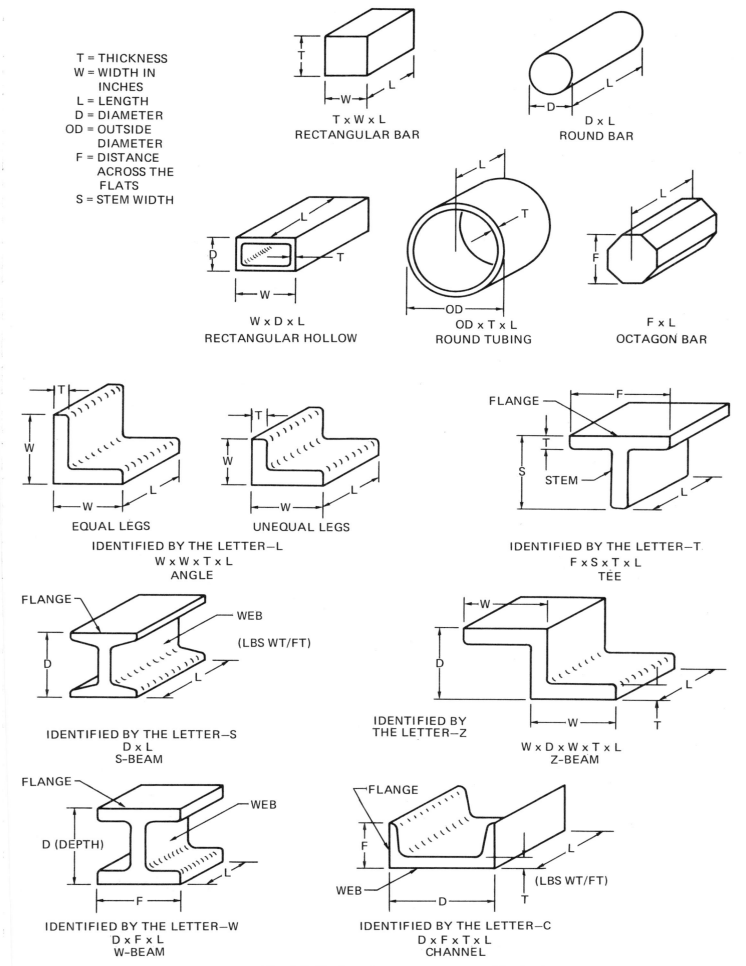

T = THICKNESS
W = WIDTH IN
    INCHES
L = LENGTH
D = DIAMETER
OD = OUTSIDE
     DIAMETER
F = DISTANCE
    ACROSS THE
    FLATS
S = STEM WIDTH

T x W x L
RECTANGULAR BAR

D x L
ROUND BAR

W x D x L
RECTANGULAR HOLLOW

OD x T x L
ROUND TUBING

F x L
OCTAGON BAR

EQUAL LEGS

UNEQUAL LEGS

IDENTIFIED BY THE LETTER—L
W x W x T x L
ANGLE

FLANGE
STEM

IDENTIFIED BY THE LETTER—T
F x S x T x L
TEE

FLANGE
WEB
(LBS WT/FT)

IDENTIFIED BY THE LETTER—S
D x L
S-BEAM

IDENTIFIED BY
THE LETTER—Z
W x D x W x T x L
Z-BEAM

FLANGE
WEB

IDENTIFIED BY THE LETTER—W
D x F x L
W-BEAM

FLANGE
WEB
(LBS WT/FT)

IDENTIFIED BY THE LETTER—C
D x F x T x L
CHANNEL

Fig. 1-2 Designating measurements of basic shapes

WEIGHTS OF MATERIALS

Figure 1-3 indicates the average weight per cubic foot of certain materials.

WEIGHTS OF MATERIALS

Material	Avg. Lbs. per Cu. Ft.	Avg. Kg. per Cu. Metre	Material	Avg. Lbs. per Cu. Ft.	Avg. Kg. per Cu. Metre
Aluminum	167.1	2676	Mahogany, Honduras, dry	35	564
Brass, cast	519	8296	Manganese	465	7448
Brass, rolled	527	8437	Masonry, granite or		
Brick, common and			limestone	165	2648
hard	125	2012	Nickel, rolled	541	8649
Bronze, copper 8, tin 1	546	8754	Oak, live, perfectly dry		
Cement, Portland, 376 lbs.			.88 to 1.02	59.3	953
net per bbl	110–115	1765–1836	Pine, white, perfectly dry	25	388
Concrete, conglomerate,			Pine, yellow, southern dry	45	706
with Portland cement	150	2400	Plastics, molded	74–137	1200–2187
Copper, cast	542	8684	Rubber, manufactured	95	1518
Copper, rolled	555	8896	Slate, granulated	95	1518
Fibre, hard	87	1377	Snow, freshly fallen	5–15	70–247
Fir, Douglas	31	494	Spruce, dry	29	459
Glass, window or plate	162	2577	Steel	489.6	7837
Gravel, round	100–125	1586–2012	Tin, cast	459	7342
Iron, cast	450	7201	Walnut, black, perfectly dry	38	600
Iron, wrought	480	7695	Water, distilled or pure rain	62.4	988
Lead, commercial	710	11,367	Zinc or spelter, cast	443	7095

Fig. 1-3 Average weights of materials

MACHINE TOOL OPERATIONS

The drafter must include all information required so a skilled machinist can manufacture a finished part from raw stock. Basic manufacturing processes shape raw stock by:

- *Cutting* into shape
- *Molding* into shape by casting or machine press
- *Pounding* into shape by forging
- *Forcing* into shape by bending
- *Fabricating* into shape, using parts manufactured from a combination of the above processes, by welding, riveting, screwing, or nailing the parts together

Machines that cut metal are called *machine tools*. There are over 400 kinds of machine tools, each designed to do a specific operation. All machine tool operations can be divided into five basic processes: Drilling, turning, planing, milling, and grinding.

A drafter does not have to be a machinist but should have a basic knowledge of machine tool operations in order to dimension drawings and "talk the language" of the skilled machinest. It is recommended that every mechanical drafting student visit a machine shop and study the five machine tool operations. A serious student should also try to take a mini-course in basic machine shop.

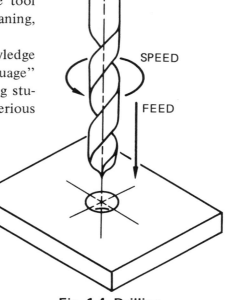

Drilling

Drilling is probably the most basic of all machine tool operations. The process is done with a rotating tool called a *drill*, figure 1-4. The drill is held by a *chuck* and is rotated and fed into the part to be drilled. Holes are drilled before boring, reaming, countersinking, or counterboring operations can be completed.

Fig. 1-4 Drilling

Turning

Turning is the process of rotating the part that is to be machined and carefully pressing a cutting tool against it as it rotates, figure 1-5. A *lathe* is used for turning down stock and can be used for other machine operations such as drilling, boring, threading, cutting, milling, grinding, and knurling. A *turret lathe* is a lathe with a six-sided tool holder, called a *turret,* to which various cutting tools are attached. This attachment enables the lathe to do many operations without resetting the tools once each of the tools in the turret has been set.

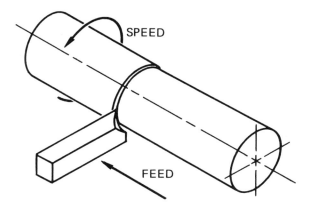

Fig. 1-5 Turning

Planing

Planing is the process of shaving material from raw stock very much like a carpenter does with a simple hand plane. The major difference is that the plane or cutting edge is stationary, and the part that is to be planed is moved back and forth, figure 1-6. A *shaper* is like a planer except, in shaping, the plane or cutting edge moves and the part is stationary. A *broach,* which is generally used to cut key slots and similar configurations, falls into the category of a planer in the way it operates. These tools can be worked horizontally, vertically, or angularly.

Grinding

Grinding is a machine operation where the part is brought into contact with a rotating abrasive wheel, figure 1-7. With this process it is possible to obtain very close, precise tolerances. Grinders that finish round parts are called *cylindrical grinders.* Those that grind flat parts are called *surface grinders.* Those that grind holes are called *internal grinders.*

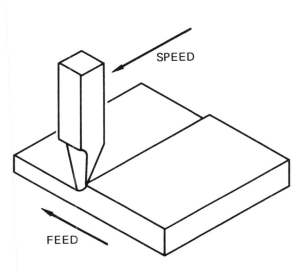

Fig. 1-6 Planing

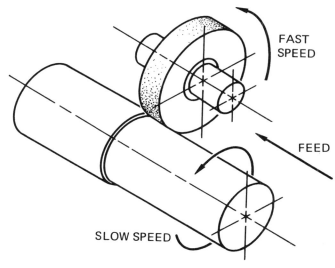

Fig. 1-7 Grinding

Milling

Milling is the process of bringing the part into contact with a rotating cutting tool having many edges, figure 1-8. The shape of the cutting edges are similar to a woodworking circular saw blade except they are usually wider and have a greater variety of shapes, figures 1-9 and 1-10. Milling machines produce cuts that are flat, round, sharp, and a combination of these shapes. Common milling machine processes are cutting slots and grooves, cutting gear teeth, making threads, boring holes, and rounding corners of parts.

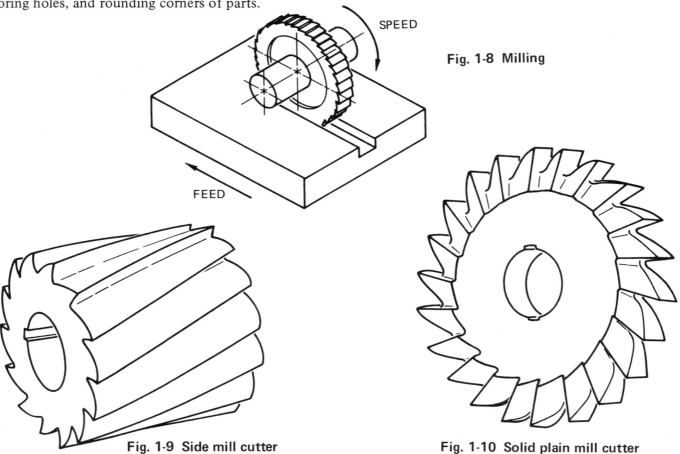

Fig. 1-8 Milling

Fig. 1-9 Side mill cutter

Fig. 1-10 Solid plain mill cutter

CASTINGS

Casting is the process of forming metal parts to rough size and shape by pouring molten metal into a mold. This process is similar to the way ice cubes are formed by pouring water into a tray and freezing it.

There are many forms of casting, varying in techniques and precision. To explain casting, one basic method – sand casting – is illustrated. Figure 1-11 shows the object to be cast.

The patternmaker constructs the pattern of the object to be cast. If the object, such as a bookend, has one flat side, the pattern is a one-piece pattern. If the object to be cast is round, a two-piece *split pattern* is used, figure 1-12.

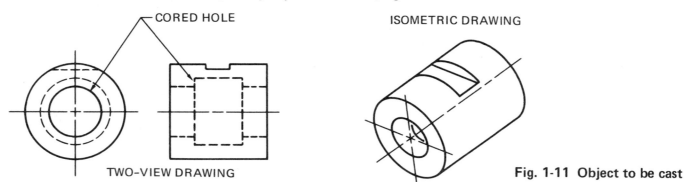

TWO-VIEW DRAWING

Fig. 1-11 Object to be cast

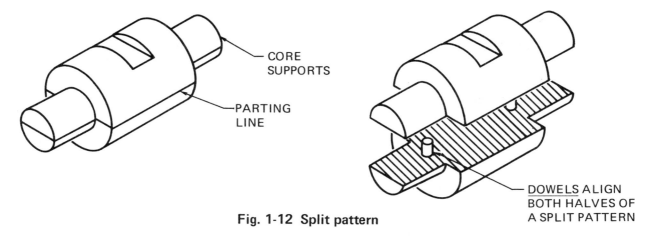

Fig. 1-12 Split pattern

CORE
SUPPORTS

PARTING
LINE

DOWELS ALIGN
BOTH HALVES OF
A SPLIT PATTERN

If the object has a large hole through it, a *core support* is located inside the mold, figure 1-13. This prevents molten metal from solidifying in the hole during the casting process. When the core is removed, a rough cored hole remains which is later machined to the correct size. No cores are used on small holes as they are simply drilled into the finished casting.

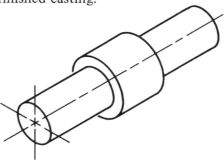

Fig. 1-13 Core made from baked sand and held together with a bonding agent

The *flask* is a hollow box with no top or bottom that holds the sand and the mold, figure 1-14. The *cope* is the top half of the flask and the *drag* is the lower half of the flask. A *socket* aligns the cope and drag. The flask is placed on a *molding board.*

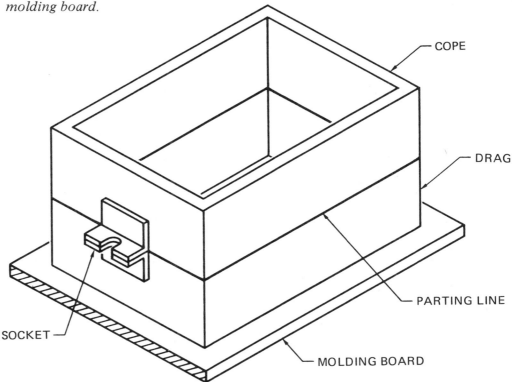

COPE

DRAG

PARTING LINE

SOCKET

MOLDING BOARD

Fig. 1-14 A flask

How to Make a Metal Casting

The following description of a sand casting is a general description only and is not intended to be used in actual casting processes.

Step 1. The lower half of the split pattern is placed on the molding board. The drag is centered around the pattern, figure 1-15.

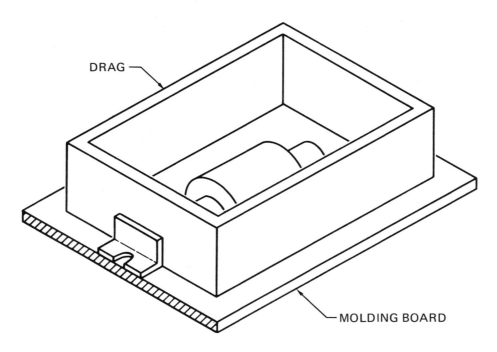

Fig. 1-15 Step 1

Step 2. The drag is filled with sifted sand which is packed firmly around the split pattern and leveled off. Another molding board is placed on top to hold the sand in place, and the drag is turned over, figure 1-16.

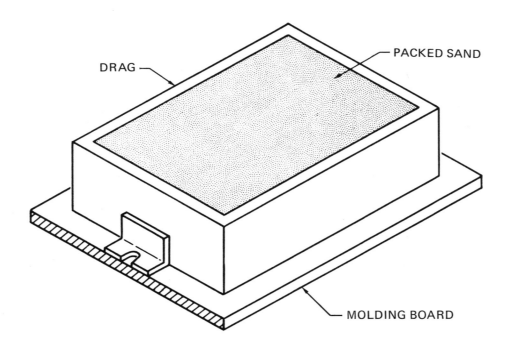

Fig. 1-16 Step 2

Step 3. The molding board is removed from the top to expose the pattern, figure 1-17.

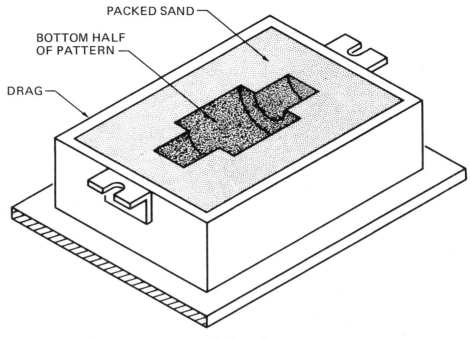

Fig. 1-17 Step 3

Step 4. The dowel pins are put in place and the top of the split pattern is positioned on the lower half, figure 1-18. The cope is then placed on the drag, locked into position, and filled with sand. The sand is tightly packed around the pattern.

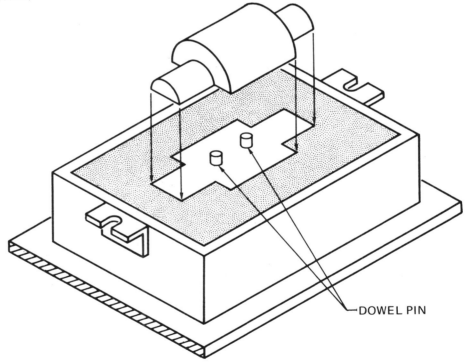

Fig. 1-18 Step 4

Step 5. Molten metal is poured into the mold through the *sprue hole.* The gases escape through the *riser hole* during the casting process. These holes are made while the pattern is in place so that the sand from the cope will not be forced into the hollow mold. An alternate method is to locate the sprue and riser holes to the left and right of the pattern and, with the flask apart, cut a groove leading from them to the mold, figure 1-19.

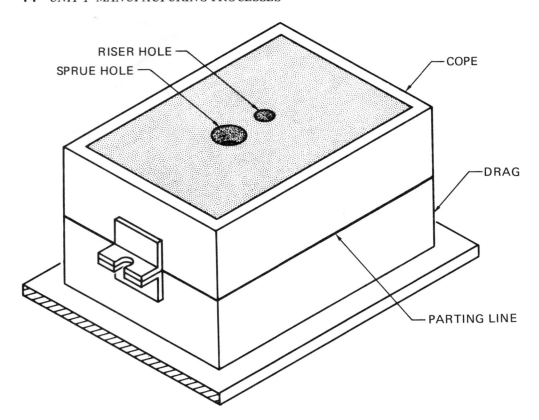

Fig. 1-19 Step 5

Step 6. After cutting the sprue and riser holes, the cope is carefully removed from the drag and set aside, and the pattern removed from the drag. This leaves the top half of the mold in the cope and the bottom half of the mold in the drag. Round patterns will lift easily from the sand. Flat patterns tend to stick and can be damaged when removed. To prevent this, flat patterns are tapered on their sides. This taper is called a *draft.* The angle of draft is shown on a casting drawing and varies with the kind of material being cast. Sharp corners on flat patterns are also rounded off slightly for the same reason.

The core is put in place and the cope returned to the drag and locked into position. The casting is now ready to be poured. After the molten metal has solidified, the casting is removed from the sand and the core is removed from the casting. The casting is now ready to be machined, figure 1-20.

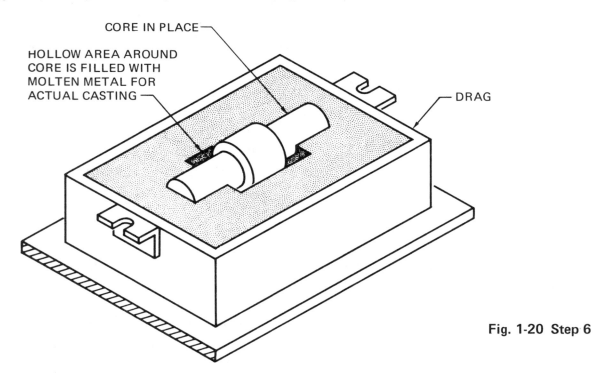

Fig. 1-20 Step 6

Section View of a Casting

A complex casting is drawn in two sections: One sh...
the other shows the machining of the casting.

Figure 1-21 is a section view of a complete flask with the c...
molten metal is poured into the funnel-shaped sprue hole until t...
cavity and comes out the riser. The riser allows air to escape wh...
filled and feeds the casting while it cools.

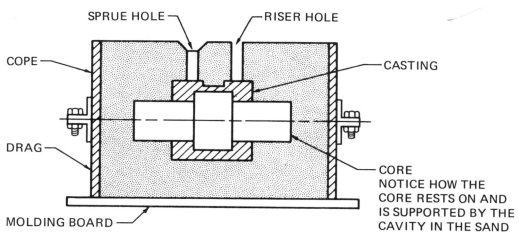

Fig. 1-21 Section view of mold ready to be cast

The sand core is easily broken up and removed, leaving a cavity inside the casting. When the casting is removed from the sand, the sprue and riser are still attached, figure 1-22.

The sprue and riser are easily removed by breaking or cutting them off, figure 1-23. They are then hand-ground. A sand casting is rough and the critical surfaces must be smoothed and machined to exact sizes.

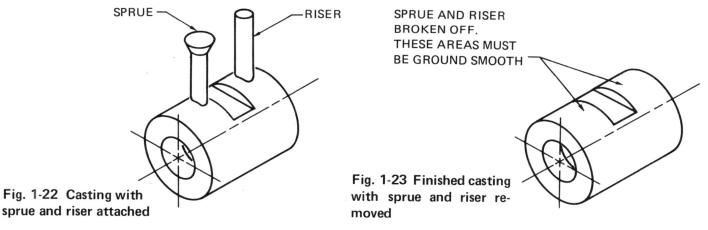

Fig. 1-22 Casting with sprue and riser attached

Fig. 1-23 Finished casting with sprue and riser removed

Rounds and Fillets

Any part that is formed by casting should be designed with rounds and fillets, figure 1-24. *Rounds* are merely rounded outside corners. *Fillets* are rounded inside corners.

Rounds and fillets are used for three reasons:

1. Greater strength for inside corners (fillets)
2. Safer to handle; no sharp corners (rounds)
3. Gives the finished casting a much better appearance (rounds/fillets)

Fig. 1-24 Rounds and fillets

ads, and Machining Lugs

Bosses and *pads* serve the same function. They are raised surfaces that are machined to provide a smooth surface for mating parts. This method of designing saves material and machining time.

A *boss* is a round, raised surface. A *pad* can be any shape raised surface. Usually the bottom surface of a part is machined to provide a smooth, solid surface to support the part. The rest of the sand casting's surface is very rough. Proper use of rounds and fillets are illustrated in figure 1-25.

A *machining lug* is an extension of a surface to be machined. It is used for holding the casting because of its shape while machining. It too is removed after machining.

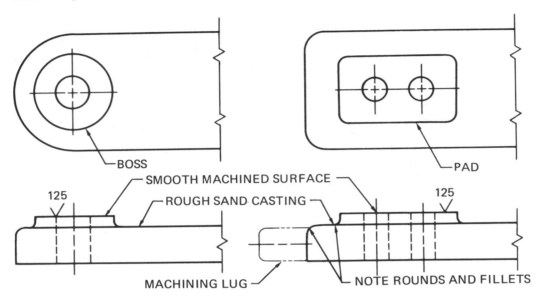

Fig. 1-25 Boss, pad, and machining lug

Ribs and Webs

Ribs and webs are similar and often confused. Think of a *rib* as a member that supports other members. A *web* simply connects various members together. Study figure 1-26. In designing castings, the general rule is to try to make all ribs and webs the same thickness. Otherwise, when the molten metal cools, thicker members cool last and tend to create warping and internal stresses.

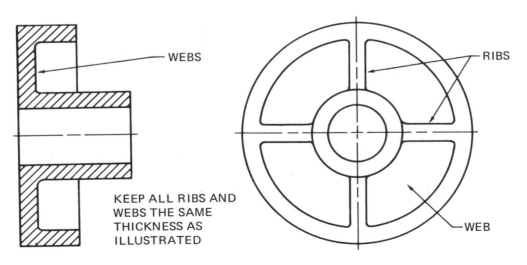

Fig. 1-26 Ribs and webs

Lug

A *lug,* sometimes referred to as an *ear,* is an extension added to the main part of the object, figure 1-27. It usually holds parts together. The thickness of the lug should be approximately the same thickness as the round body thickness.

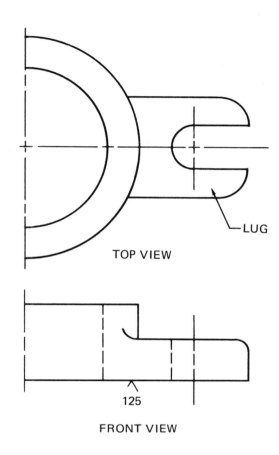

TOP VIEW

FRONT VIEW

Fig. 1-27 Lugs

Shrink Rule

Patterns must be made larger than the desired size of the casting to compensate for the shrinkage which occurs when metal cools. A *shrink rule* is used by the patternmaker to overcome this difference. Different shrink rules are used for different materials.

Shrinkage allowances have been established for various metals:

- Cast iron and malleable iron — 1/8 inch per foot (10 mm per meter)
- Copper, aluminum, and bronze — 3/16 inch per foot (16 mm per meter)
- Steel — 1/4 inch per foot (21 mm per meter)
- Lead 5/16 inch per foot (26 mm per meter)

The patternmaker uses the shrink-rule measurement which correspond to the shrinkage allowance. In this way the pattern will be large enough to compensate for shrinkage, and the final casting will shrink to the desired original size.

UNIT REVIEW

45-minute time limit

1. List three kinds of heat-treating methods and briefly explain each.

2. Why is an alloy used in place of a pure metal?

3. Explain the difference between ferrous and nonferrous metal.

4. What is fatigue limit?

5. Explain case hardening.

6. In a SAE 2010 steel callout, what do the first two numbers (20) indicate? The last two numbers (10)?

7. What are the five basic machine tool operations?

8. What does a chuck do?

9. What are the five manufacturing processes used to shape raw stock?

10. Explain shrinkage allowance.

11. What is the difference between a pad and a boss?

12. List three functions of a riser.

13. What is the basic rule for designing ribs and webs?

14. What makes up the flask in the casting process?

Before proceeding to the next unit:

_____ Instructor's approval

_____ Progress plotted

UNIT 2

BASIC WELDING

OBJECTIVE

The student will study various welding methods and learn how to use welding symbols on engineering drawings.

PRETEST

30-minute time limit

Questions

1. What is a fusion weld?

2. What is a resistance weld?

3. What does it mean when a weld symbol is placed below the reference line? Above the reference line?

Identification

Carefully add the correct weld symbol for each illustration. Place the symbol above or below the reference line per drafting standards.

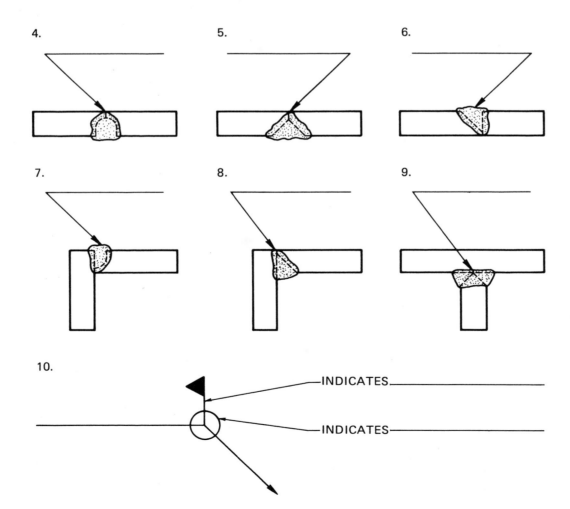

4.

5.

6.

7.

8.

9.

10.

INDICATES _____

INDICATES _____

RELATED TERMS

Give a brief definition of each term as progress is made through the unit.

Fusion welding _____

Resistance welding _____

Types of joints _____

Reference line _____

Weld symbols _____

Pitch _____

Fusion Welding

In *fusion welding*, a welding rod is melted and combined with the metal parts that are to be fastened together. The parts will be permanently joined after cooling. The process can be done using torches or high electric power.

Types of Typical Joints

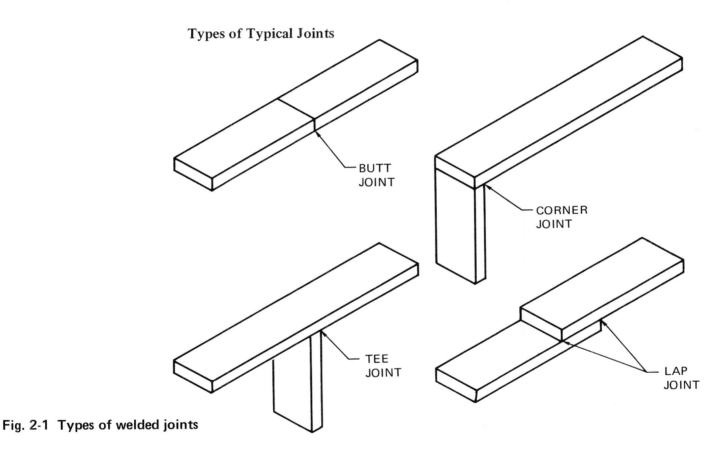

Fig. 2-1 Types of welded joints

Symbols

Figure 2-2 indicates the symbols for each type of weld. Figure 2-3 gives an example of each type of weld.

TYPE OF WELDS

① BACK OR BACKING WELD	② FILLET WELD	③ PLUG WELD	④ SQUARE WELD	⑤ V WELD	⑥ BEVEL WELD	⑦ U WELD	⑧ J WELD
⌒	◺	▭	‖	⋁	⋁	∪	⋃

Fig. 2-2 Weld symbols

Figure 2-3 gives an example of each type of weld.

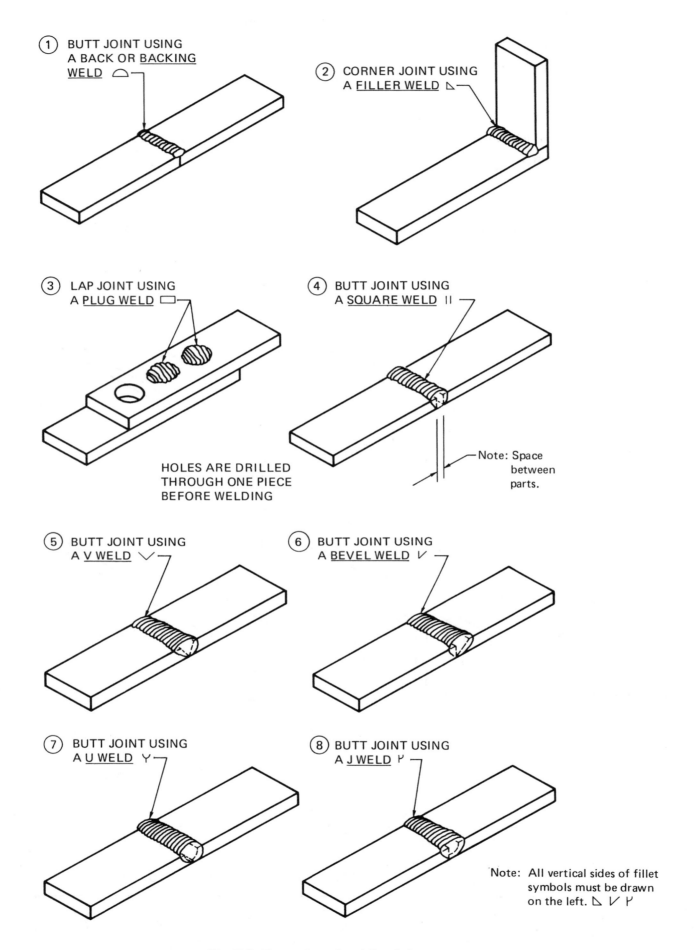

① BUTT JOINT USING
A BACK OR <u>BACKING</u>
<u>WELD</u> ◠

② CORNER JOINT USING
A <u>FILLER WELD</u> ◿

③ LAP JOINT USING
A <u>PLUG WELD</u> ▭

④ BUTT JOINT USING
A <u>SQUARE WELD</u> ‖

HOLES ARE DRILLED
THROUGH ONE PIECE
BEFORE WELDING

Note: Space
between
parts.

⑤ BUTT JOINT USING
A <u>V WELD</u> ⋁

⑥ BUTT JOINT USING
A <u>BEVEL WELD</u> ⋁

⑦ BUTT JOINT USING
A <u>U WELD</u> ⋃

⑧ BUTT JOINT USING
A <u>J WELD</u> ⌐

Note: All vertical sides of fillet
symbols must be drawn
on the left. ◿ ⋁ ⌐

Fig. 2-3 Examples of welding joints

Placing Weld Symbols

Rule 1. When the weld symbol is placed below the reference line, figure 2-4(B), the weld appears on the same side as the arrowhead.

Rule 2. When the weld symbol is placed above the reference line (A), the weld appears on the opposite side of the arrowhead.

Rule 3. When the weld symbol is placed above and below the reference line (A and B), the weld appears on both sides of the arrowhead.

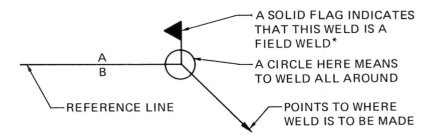

A SOLID FLAG INDICATES THAT THIS WELD IS A FIELD WELD*

A CIRCLE HERE MEANS TO WELD ALL AROUND

REFERENCE LINE

POINTS TO WHERE WELD IS TO BE MADE

Fig. 2-4 Placing standard welding symbols

Figure 2-5 shows examples of rules 1, 2, and 3 using a filler weld symbol.

Usually material over .125 inch (3) thick requires a *groove* (square, V, beveled, U, or J). Using the basic weld symbol, point the arrowhead toward the part that has the groove, figure 2-6.

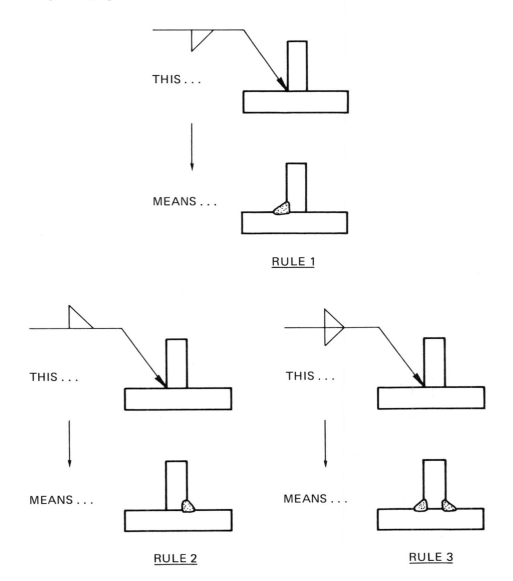

THIS . . .

MEANS . . .

RULE 1

THIS . . .

MEANS . . .

RULE 2

THIS . . .

MEANS . . .

RULE 3

Fig. 2-5 Welding reference line rules

*A field weld symbol indicates that the weld must be made at the work site and not in the welding shop.

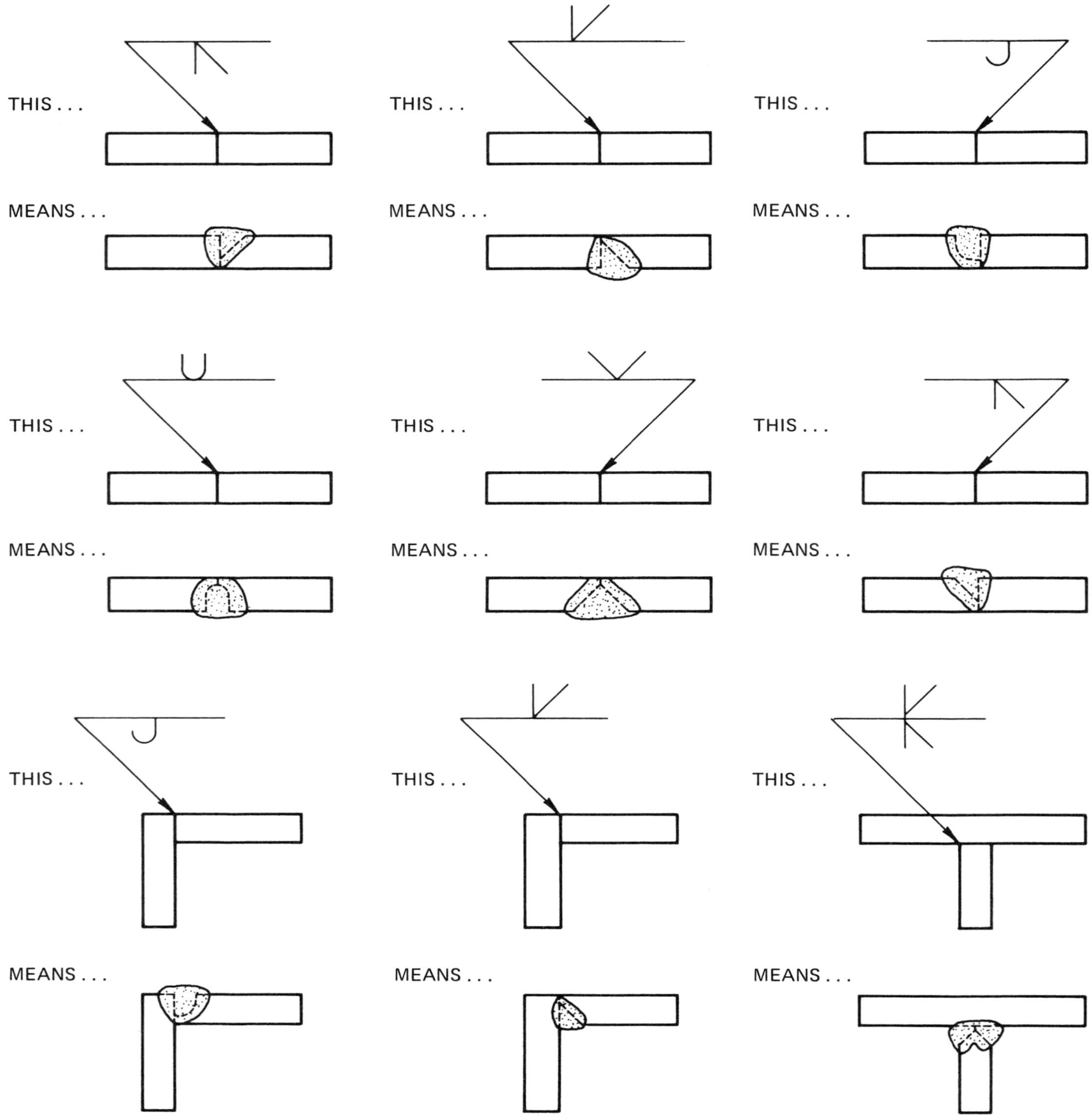

THIS . . .

MEANS . . .

THIS . . .

MEANS . . .

THIS . . .

MEANS . . .

THIS . . .

MEANS . . .

THIS . . .

MEANS . . .

THIS . . .

MEANS . . .

THIS . . .

MEANS . . .

THIS . . .

MEANS . . .

THIS . . .

MEANS . . .

Groove welds shown require that the groove be machined to a specific size conforming to a specific delineation.

Fig. 2-6 Grooves in welding symbols

Practice Exercise 2-1

Below each weld callout, sketch the position and shape of the weld called for. Compare your work to the answers on page 35.

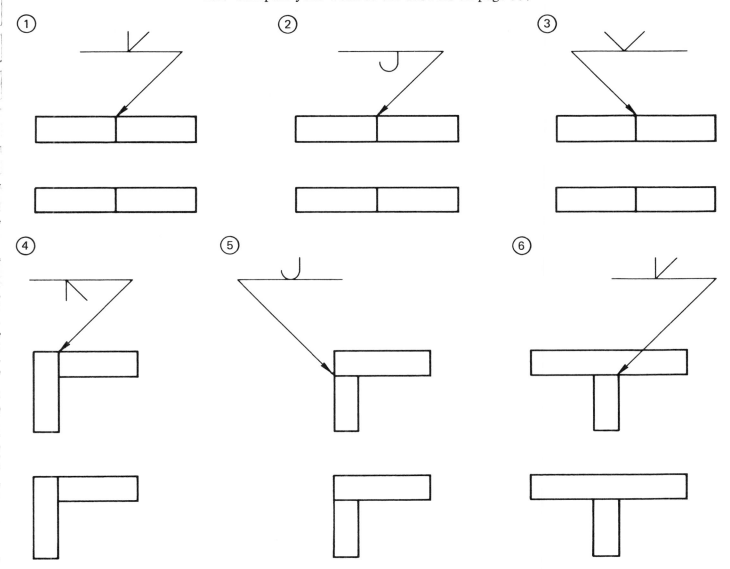

Reference Line Notations

There are various *notations* placed on or around the reference line. Figure 2-7 lists a few of the more widely used standard notations. Each tells the welder exactly how the drafter wants the part (s) welded. The .25" X .375 (6 X 10) *size notation* means the weld is approximately .25" X .375 and is welded the whole length of the part.

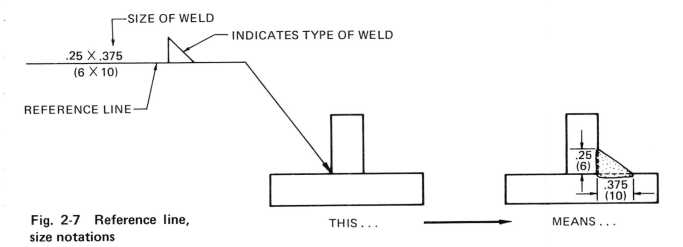

Fig. 2-7 Reference line, size notations

In figure 2-8, two more notations are added. The first means the length of each weld, and the second means the distance from center to center of each weld or pitch. *Pitch* refers to the distance, center to center, of each weld. The 2-4 notation means that the weld is to be 2 inches long with a center-to-center distance of 4 inches. The notations would indicate millimetres if the metric system is used.

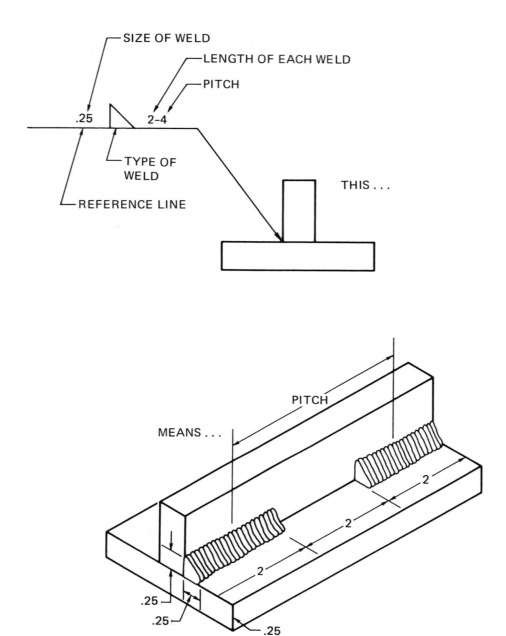

Fig. 2-8 Reference line, pitch notation

Practice Exercise 2-2

On the reference line, draw the welding symbol required to obtain the weld and spacing shown on the isometric drawings given. Compare your work to the answers on page 35.

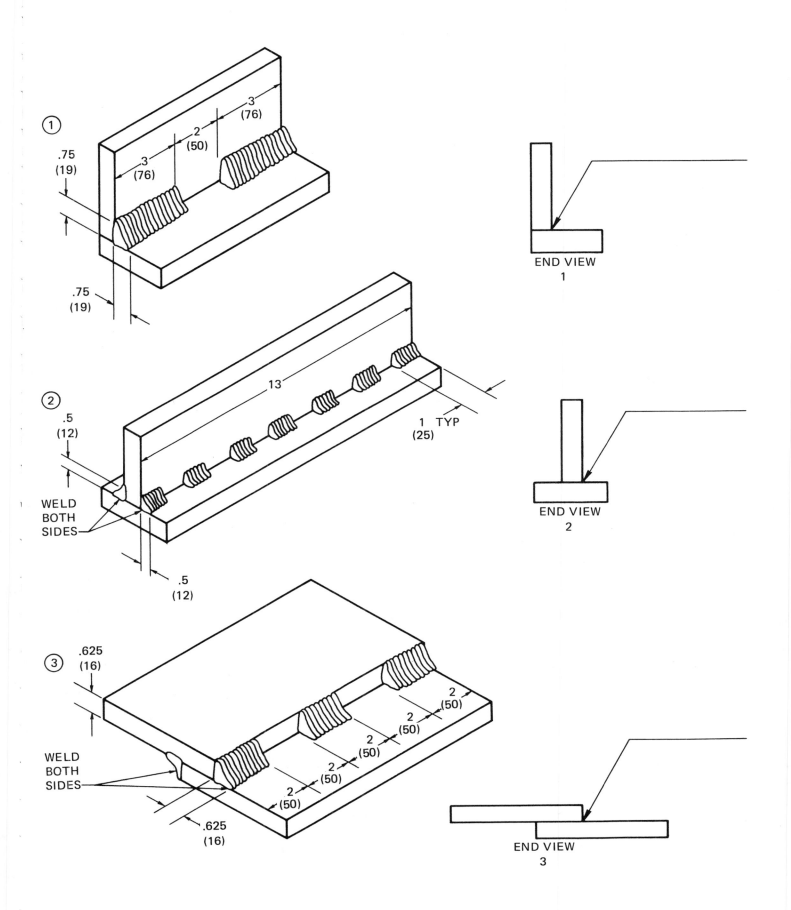

RESISTANCE WELDING

Resistance welding is the process of passing an electric current through a spot where the parts are to be joined. Symbols for resistance welding are shown in figure 2-9.

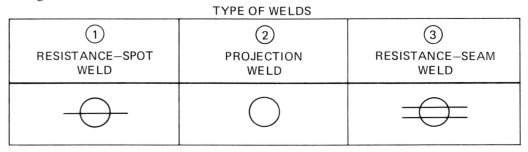

Fig. 2-9 Symbols for resistance welds

Study figure 2-10. Note that the symbols are similar to those used in fusion welding. As with fusion welding, notations are done on the side of the arrow where they appear.

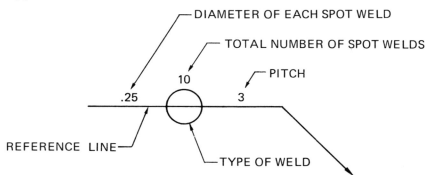

Fig. 2-10 Reference line for resistance welding

Spot Welding

Spot welding joins parts together with small circles or spots of heat. Figure 2-11 shows how a drafter would draw and dimension a drawing. Figure 2-12 shows how a welder would spot weld the project.

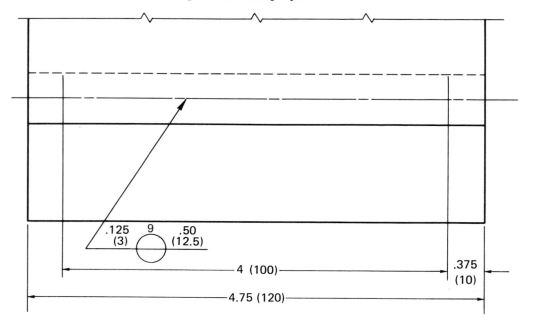

Fig. 2-11

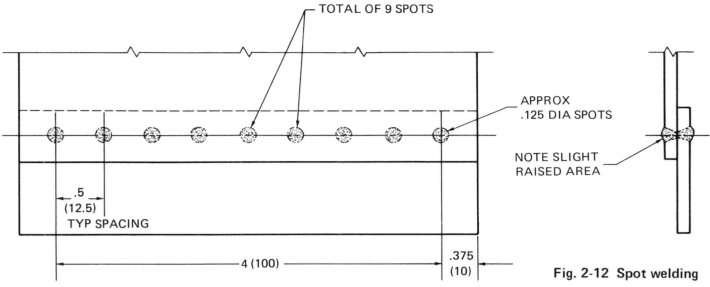

Fig. 2-12 Spot welding

Flush Symbol

A *flush symbol* is used to indicate that one or both surfaces must be ground smooth, figure 2-13.

Fig. 2-13 Flush symbol added to welding symbol

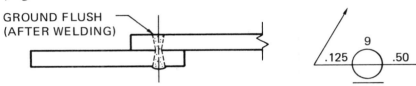

Projection Weld

A *projection weld* is the same as a spot weld except one part has a *dimple* stamped into it at each spot where it is to be welded. This dimple allows more penetration and, as a result, is a better weld, figure 2-14.

Fig. 2-14 Dimple symbol and application

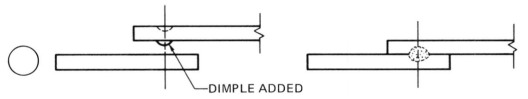

Figure 2-15 shows how a drafter would draw and dimension a drawing. Note that the symbol is located below the reference line indicating that the dimple is on the part that is on the arrow side.

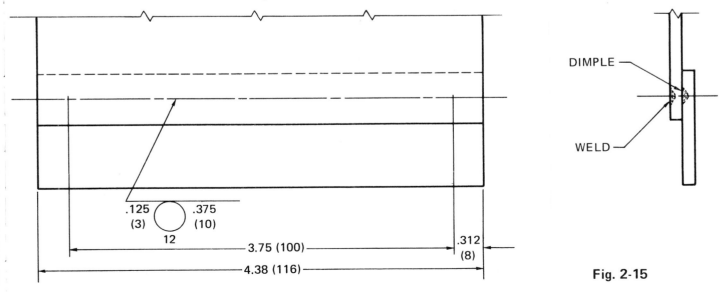

Fig. 2-15

Figure 2-16 shows how the welder would weld the project.

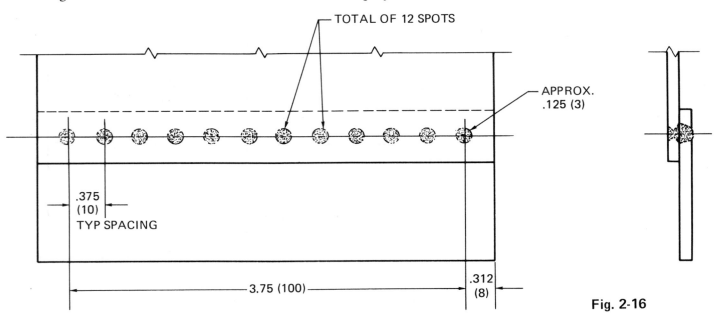

Fig. 2-16

Resistance Seam Weld

A *resistance seam weld* is like the spot weld process except the weld is continuous from start to finish. Figure 2-17 shows how the drafter would draw and dimension a drawing.

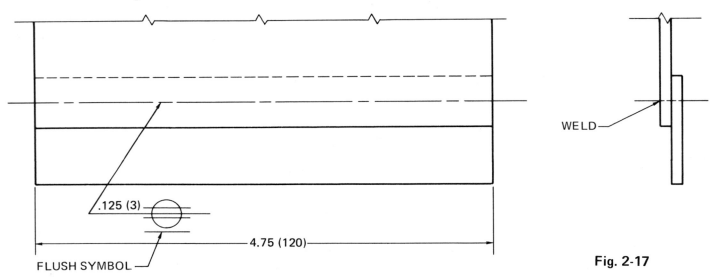

Fig. 2-17

Figure 2-18 shows how the welder would seam weld the project.

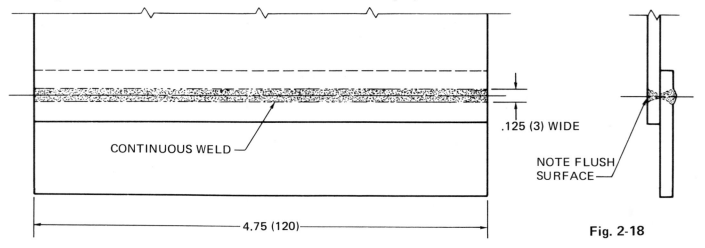

Fig. 2-18

Practice Exercise 2-3

Study each isometric sketch. On the reference lines, place the information needed to obtain the desired welds illustrated at the left of each problem. Compare your work to the answers on page 35.

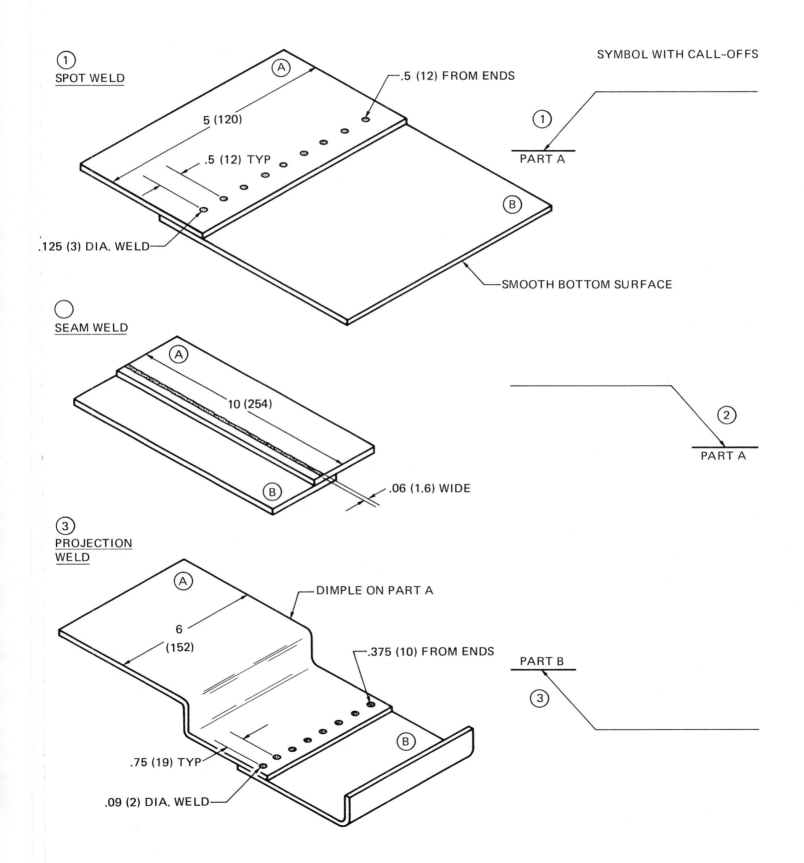

SYMBOL WITH CALL-OFFS

① SPOT WELD

.5 (12) FROM ENDS

5 (120)

.5 (12) TYP

.125 (3) DIA. WELD

SMOOTH BOTTOM SURFACE

PART A

○ SEAM WELD

10 (254)

.06 (1.6) WIDE

②

PART A

③ PROJECTION WELD

DIMPLE ON PART A

6 (152)

.375 (10) FROM ENDS

.75 (19) TYP

.09 (2) DIA. WELD

PART B

③

Practice Exercise 2-4

Add the correct filler weld symbol, size, length, and pitch to the blank reference lines. Place the symbols on the correct side of the reference line. Take approximate measurements directly from the three-view drawing. Compare your work to the answer on page 36.

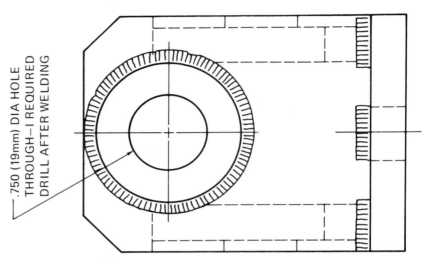

.750 (19mm) DIA HOLE THROUGH—I REQUIRED DRILL AFTER WELDING

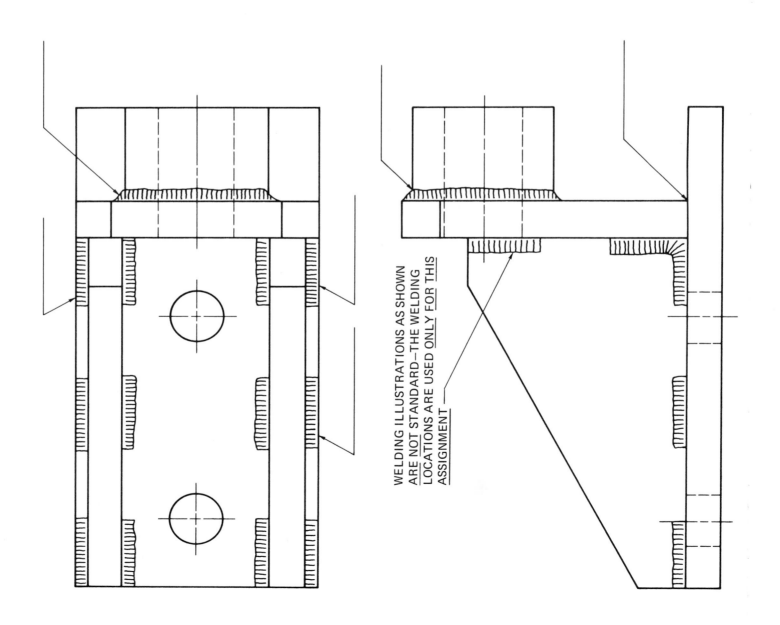

WELDING ILLUSTRATIONS AS SHOWN ARE NOT STANDARD—THE WELDING LOCATIONS ARE USED ONLY FOR THIS ASSIGNMENT

ANSWERS TO PRACTICE EXERCISES

Carefully compare your work to the answer for each exercise. Refer any questions to the instructor.

Exercise 2-1

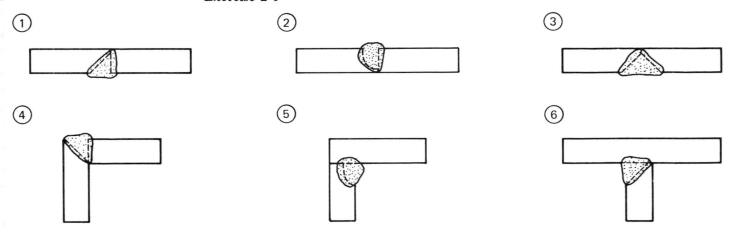

Exercise 2-2

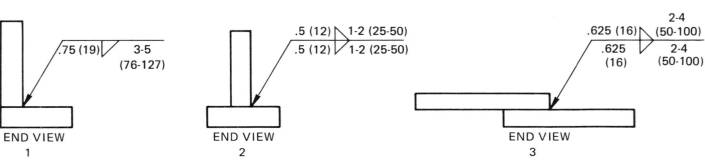

Exercise 2-3

1. .125″ (3) = diameter of each spot weld
   9 = number of welds
   .50″ (12) = pitch or how far apart each spot weld is
Note: The flush mark above the reference line indicates that the finish is smooth on the opposite side of the arrowhead.

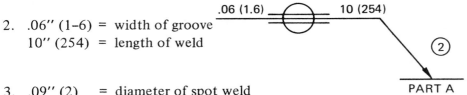

2. .06″ (1–6) = width of groove
   10″ (254) = length of weld

3. .09″ (2) = diameter of spot weld
   8 = number of welds
   .75″ (19) = pitch or how far apart each spot is
Note: The weld symbol is above the reference line indicating that the dimples are on Part 'A'.

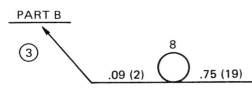

Exercise 2-4

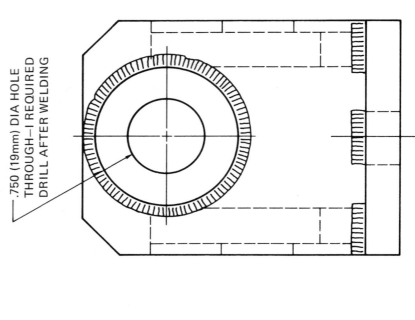

.750 (19mm) DIA HOLE THROUGH—I REQUIRED DRILL AFTER WELDING

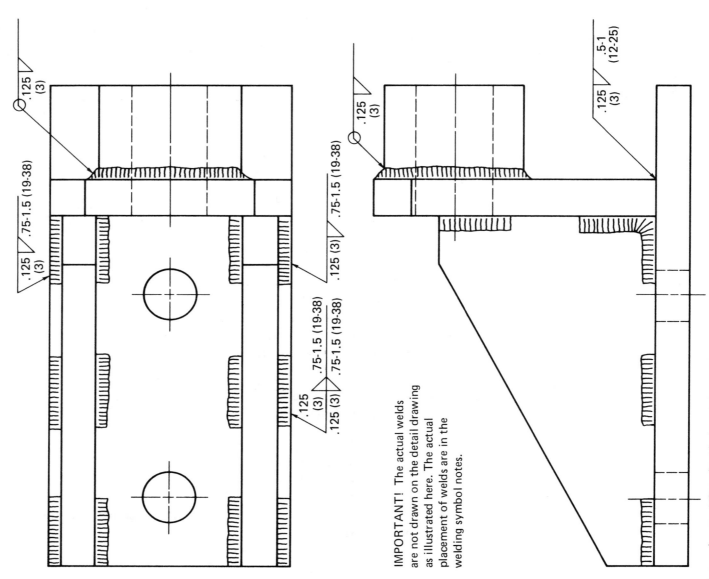

.125 (3)

.75-1.5 (19-38)

.125 (3)

.125 (3)

.75-1.5 (19-38)

.75-1.5 (19-38)

.125 (3)

.125 (3)

.125 (3)

.75-1.5 (19-38)

.5-1 (12-25)

.125 (3)

IMPORTANT! The actual welds are not drawn on the detail drawing as illustrated here. The actual placement of welds are in the welding symbol notes.

UNIT REVIEW

Problems

10-minute time limit

1. Add extension line, arrowhead, and welding symbol to the three drawings located below the welds.

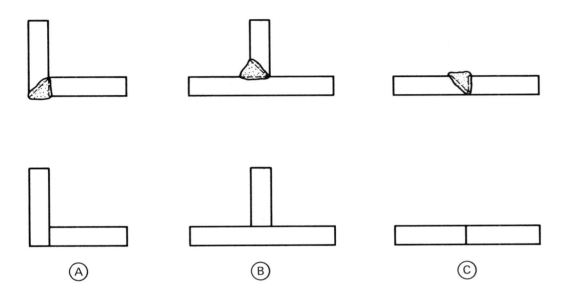

2. Add complete welding symbol, with all notations, to the reference line given.

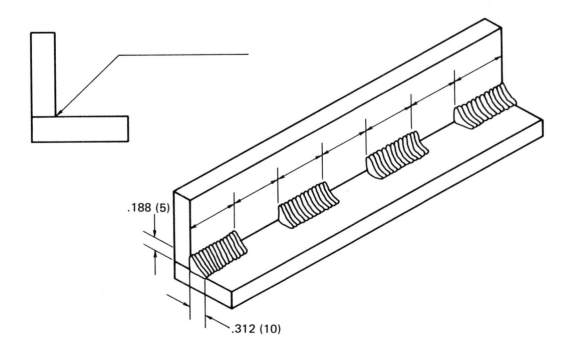

.188 (5)

.312 (10)

3. Explain fully what a, b, c, and d mean.   Place the answers in the spaces
    provided.

    a =

    b =

    c =

    d =

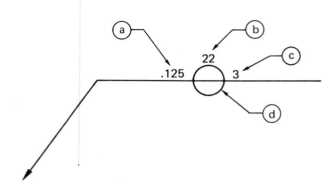

Before proceeding to the next unit:

_____ Instructor's approval

_____ Progress plotted

UNIT 3

FASTENERS

OBJECTIVE

The student will learn how to draw various fasteners using dimension charts.

PRETEST

60-minute time limit

1. List the head size of a 5/8-inch socket-head cap screw.

2. List the six standard points used on a setscrew.

3. What is the distance across the flats of a 7/8-inch square nut?

4. What size hole is recommended for a 1/4-inch cotter pin?

5. What size diameter screw is considered a machine screw? What size range do cap screws come in?

6. If you have a 7/16-20 UNF threaded screw and rotate it ten full turns, how far will the end travel?

7. What is the standard length of threads for a 1/2-inch diameter slotted-head cap screw that is 2 1/2 inches long?

8. What is the tolerance on a square key with a 2-inch diameter shaft using key stock?

9. What is the minimum thread depth recommended for use in cast iron?

10. What is the thread depth of a 9/16-inch coarse thread?

RELATED TERMS

Give a brief definition of each term as progress is made through the unit.

Permanent fasteners _____

Temporary fasteners _____

Screw threads _____

Crest _____

Root _____

Pitch _____

Depth of thread _____

Thread angle _____

Thread forms _____

Tap _____

Die _____

Blind hole _____

Fastener callout _____

T.P.I. _____

Machine screw _____

Cap screw _____

Setscrew _____

Points _____

Thread relief _____

FASTENERS

Objects that are assembled must be held together with some type of fastener or by a fastening procedure. There are two major classifications of fasteners, permanent and temporary. *Permanent fasteners* are used when parts will not be disassembled. *Temporary fasteners* are used when the parts will be disassembled at some future time.

Permanent fastening methods include:

- Welding
- Brazing
- Stapling
- Nailing
- Gluing
- Riveting

Temporary fasteners include:

- Screws
- Bolts
- Keys
- Pins

There are many types and sizes of fasteners, each designed for a particular function. Welding procedures were discussed in Unit 2. Unit 3 provides detailed information on screw threads and the more common temporary fasteners. Although screw threads have other important uses, such as adjusting parts and transmitting power, only their use as a fastener will be discussed.

THREADS

Figure 3-1 illustrates a Unified National thread and labels the terms associated with all threads.

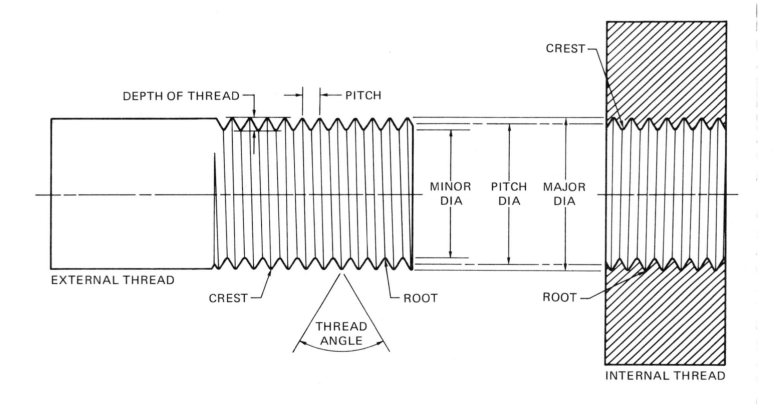

Fig. 3-1 Unified National thread

Common Thread Forms

Figure 3-2 illustrates the most common type of thread forms:
- Unified National
- Square — transmits power
- ACME — transmits power
- Buttress — transmits power in only one direction
- Worm — transmits power

Note that pitch (P) determines all other dimensions.

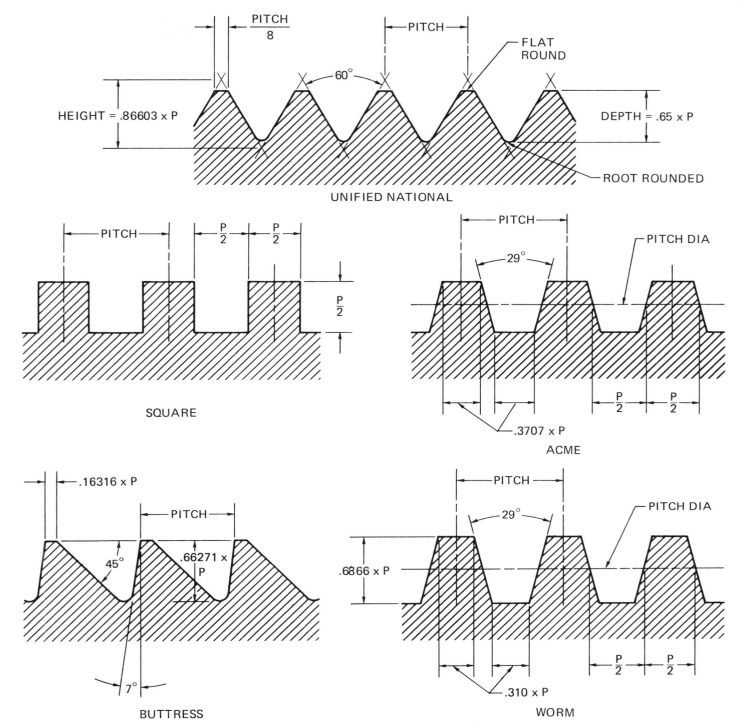

Fig. 3-2 Common thread forms

Classes of Fits

The fit of a thread is the amount of play between the screw and the nut when they are assembled together. There are three classes of fits for external threads and internal threads, figure 3-3.

CLASS OF FIT		CHARACTERISTICS
External Thread	Internal Thread	
1A	1B	Loosest fit. Used where easy assembly and disassembly is important.
2A	2B	Average fit. Used for ordinary fasteners.
3A	3B	Tight fit. Used when a snug fit is required for greater precision.

Fig. 3-3 Class of fit

TAP AND DIE

There are various methods to produce inside and outside threads. The simplest method uses thread cutting tools called *taps* and *dies*. A tap cuts internal threads, while a die cuts external threads, figures 3-4 and 3-5.

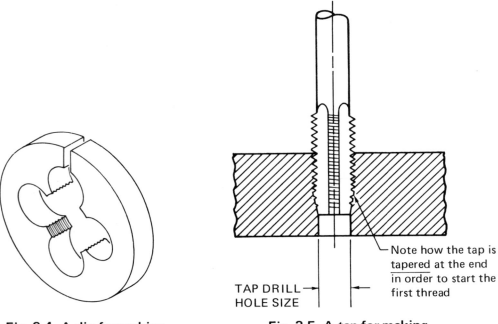

TAP DRILL
HOLE SIZE

Note how the tap is tapered at the end in order to start the first thread

Fig. 3-4 A die for making external threads

Fig. 3-5 A tap for making internal threads

Single and Multiple Threads

The *single thread* is standard. *Double* and *triple* threads are used when speed or travel distance is important, figure 3-6. The single thread, however, has more holding power.

A good example of double or triple threads is an inexpensive ball-point pen. Carefully take a ball-point pen apart and study the end of the external threads that hold the parts together. Note how fast the parts screw together. This is a characteristic of multiple threads.

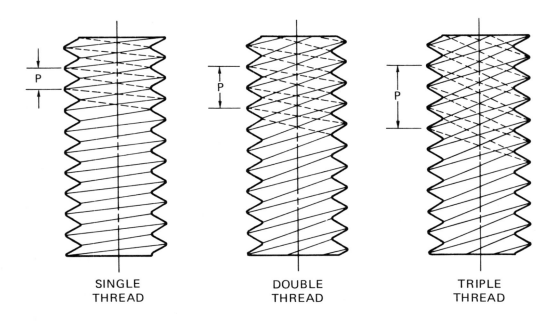

SINGLE
THREAD

DOUBLE
THREAD

TRIPLE
THREAD

Fig. 3-6 Single and multiple threads

Thread Representation

Figure 3-7 shows a normal view of an *external thread* made with a threading die. Figure 3-8 is a representative drawing showing the information needed to manufacture the thread. The dash lines used to draw the minor diameter are not hidden edge lines but merely indicate the *depth of thread*.

Figure 3-9 shows a normal view of an *internal thread* made with a tap. Figure 3-10 shows the representative drawing made by the drafter. In the representative drawing, the minor diameter is made the same size as the diameter of the tap drill recommended. The major diameter in the drawing is the same diameter as the major diameter of the thread being represented.

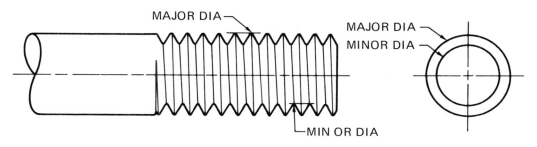

Fig. 3-7 External thread

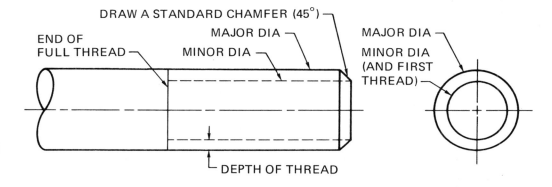

Fig. 3-8 External thread representation

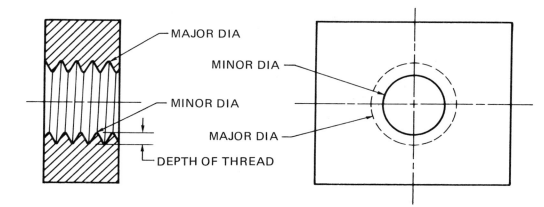

Fig. 3-9 Internal thread

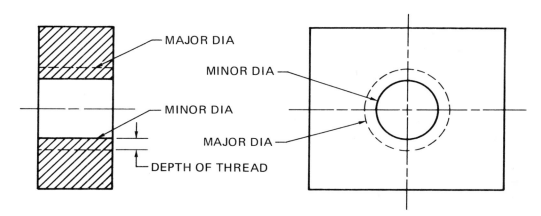

Fig. 3-10 Internal thread representation

Threaded Blind Hole

A *blind hole* is a hole drilled only part way through a piece of stock. Blind holes are often tapped to produce internal threads for screws. Figure 3-11 shows how blind holes are represented on drawings. The representative drawing is done to make sure the tap enters the stock deeply enough to assure the number of threads needed.

1. A hole is first drilled with a tap drill.
2. The tap is then turned into the tap hole.
3. The tap cuts threads as it enters the tap hole. Because the tap is tapered, it does not cut threads all the way to the bottom of the tap hole.

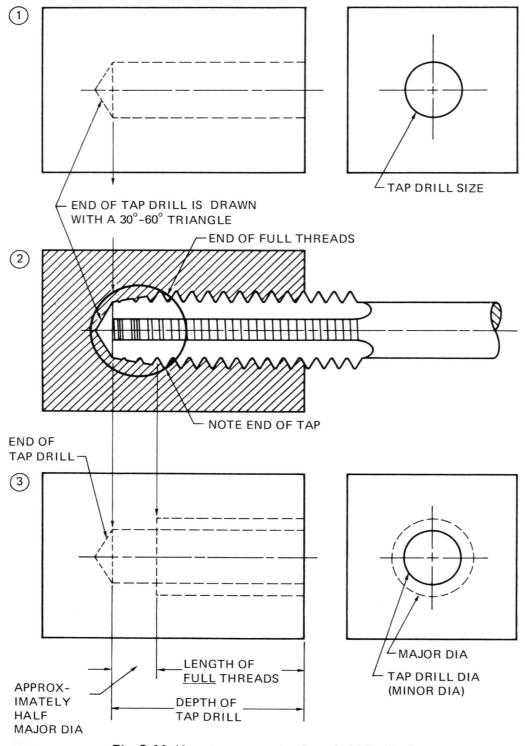

Fig. 3-11 How to represent a threaded blind hole

FASTENER CALLOUTS

Although all companies have not yet adopted the same standard for callouts, it is important that all workers within a company use the same method. The callout in figure 3-12 is the national standard callout.

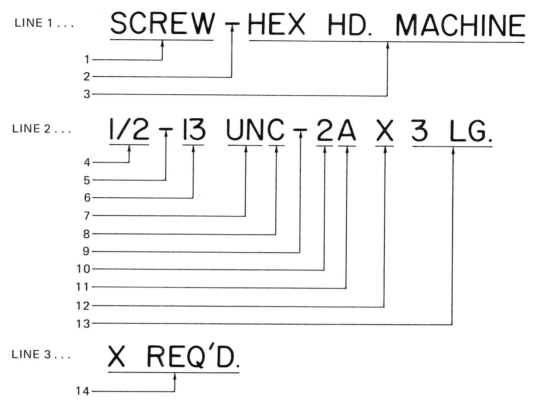

Fig. 3-12 Standard fastener callout

Line 1

1. Type of fastener
2. Dash
3. Description of fastener

Line 2

4. Nominal size
5. Dash
6. T.P.I. (threads per inch)
7. Thread series (Unified National)
8. Thread coarseness (F = fine, EF = extra fine, C = coarse)
9. Dash
10. Class of fit (1 = loose, 2 = average, 3 = tight)
11. External (A) *or* internal (B) thread
12. X = separation
13. Length (if required)

Line 3

14. Number of fasteners required

Examples of Fastener Callouts:

- NUT – HEX HD.
  5/8 – 18 UNF
  2 REQ'D
- WASHER – LOCK
  3/8
  1 REQ'D

- PIN – COTTER
  1/16 x 1 LG.
  10 REQ'D

At this time the United States, Canada, and the United Kingdon use the inch series of *Unified Thread Form.* This system designates the diameter and number of threads per inch along with a suffix indicating the thread series.

Example: 1/4 – 20 UNC

Threads are designated in the metric system using a similar approach, figure 3-13.

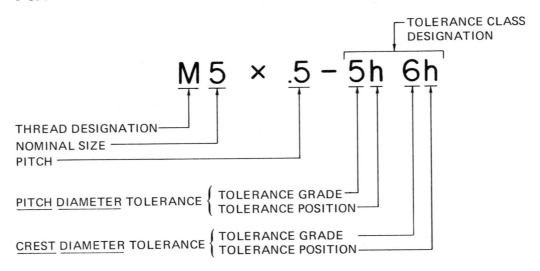

Fig. 3-13 Standard metric fastener callout

M = Shows the thread to be a metric thread

5 = Nominal size diameter in millimetres

X = Separation

.5 = Pitch of thread is 0.5 millimetres

– = Dash

5h = Tolerance class of pitch diameter

6h = Tolerance class of crest diameter

In the metric thread callout, a lower case h or g indicates *external threads,* while an upper H or G indicates *internal threads.*

In figure 3-13, the last two number and letter combinations refer to the *tolerance grades* and position of the *pitch diameter* and *crest diameter.* They are similar to the class of fit shown on line 10 in figure 3-12 dealing with threads in the inch system of callouts.

The diagram in figure 3-14 explains the symbols giving *class of fit* in the metric system.

Class of Fit	Tolerance Class	
	Bolts & Screws	Nuts
Fine	4h	5H
Medium	6g	6H
Coarse	8g	7H

Fig. 3-14 Class of fit — metric system

Metric thread callouts for fine threads use the prefix M, the diameter, and the pitch. When calling out coarse threads, only the prefix M and the diameter is used. The metric system uses coarse threads most frequently. Because the terminology of the inch system and the metric system is so different, refer to this text before adding any metric callouts to a drawing.

MEASURING THREADS

One method of measuring *threads per inch* (T.P.I.) is to place a standard scale on the crests of the threads and count the number of full threads within one inch of the scale, figure 3-15. If only part of an inch of stock is threaded, count the number of full threads in one-half inch and multiply by two to determine T.P.I.

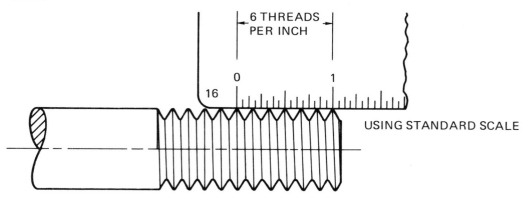

Fig. 3-15 Measuring threads with a standard scale

A simple, more accurate method of determining threads per inch is to use a *screw thread pitch gauge,* figures 3-16 and 3-17. By trial and error the various fingers or leaves of the gauge are placed over the threads until one is found that fits exactly into the threads. Threads per inch are then read directly on each leaf of the gauge.

Appendix B gives a dimension and size chart for the American National Standard Unified and American National Thread Series.

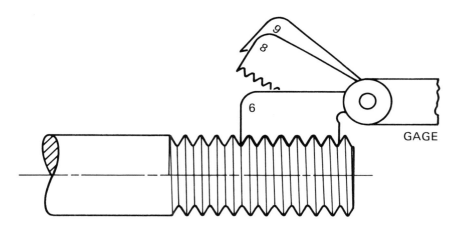

Fig. 3-16 Measuring threads using a screw thread pitch gauge

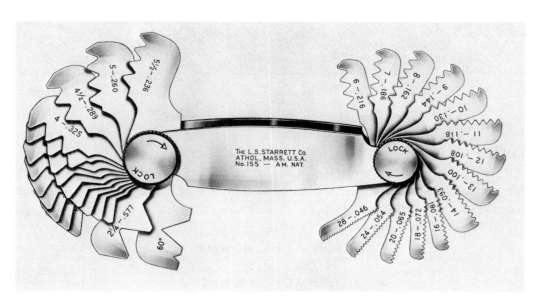

Fig. 3-17 Screw thread pitch gauge

RIGHT-HAND AND LEFT-HAND THREADS

Threads can be either right-handed or left-handed. In order to distinguish between a right-hand or left-hand thread, use this simple trick.

A right-hand thread winding tends to lean to the left, figure 3-18. If the thread leans to the left, the right-hand thumb points in the same direction. If the thread leans to the right, figure 3-19, the left-hand thumb leans in that direction indicating that it is a left-hand thread.

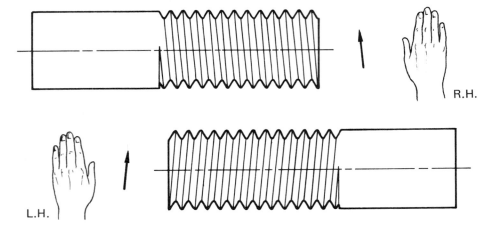

Fig. 3-18 Right-hand thread

Fig. 3-19 Left-hand thread

MACHINE SCREWS

Machine screws use a nut to fasten parts together. They also may be used in a tapped hole, figure 3-20. The five standard head types are flat, round, oval, fillister, and pan. They range in numbered sizes from #0 to #12 and in fractional sizes from 1/4 inch to 3/4 inch in length.

A typical callout for a machine screw is:

SCREW – RD. HD. MACH.

This is read as:

ROUND HEAD MACHINE SCREW

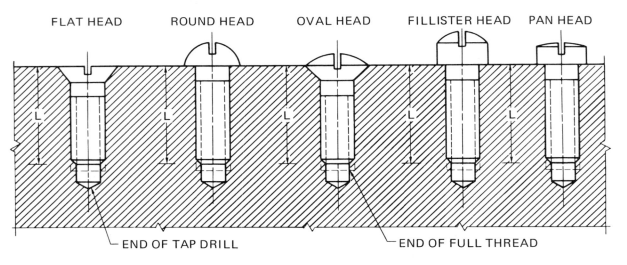

Fig. 3-20 Machine screws

CAP SCREWS

There are five standard types of heads under the classification of *cap screws:* flat, round, fillister, hex, and socket head, figure 3-21. The cap screw is larger than the machine screw and starts at 1/4 inch size through 1 1/4 inch size.

A typical callout for a cap screw is:

SCREW – HEX. HD. CAP

This reads as:

HEX HEAD CAP SCREW

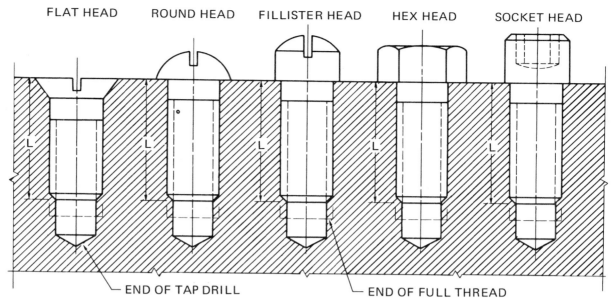

FLAT HEAD   ROUND HEAD   FILLISTER HEAD   HEX HEAD   SOCKET HEAD

END OF TAP DRILL   END OF FULL THREAD

Fig. 3-21 Cap screws

FLAT-HEAD SCREW

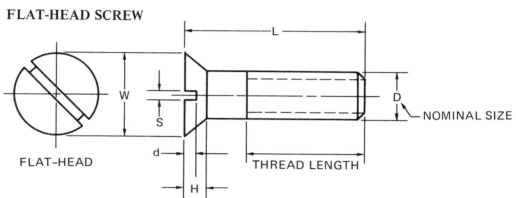

FLAT-HEAD

Fig. 3-22 Flat-head screw representation

Type	Nom. Size (Inch)	– D –		– W –	– H –	– S –	– d –
		Inch	mm	Inch	Inch	Inch	Inch
Mach.	0	.060	1.524	.119	.035	.023	.015
	1	.073	1.854	.146	.043	.026	.019
	2	.086	2.108	.172	.051	.031	.023
	3	.099	2.514	.199	.059	.035	.027
	4	.112	2.845	.225	.067	.039	.030
	5	.125	3.175	.252	.075	.043	.034
	6	.138	3.505	.279	.083	.048	.033
	8	.164	4.166	.332	.100	.054	.045
	10	.190	4.826	.385	.116	.060	.053
	12	.216	5.486	.438	.132	.067	.060
Cap	1/4	.250	6.350	.507	.153	.075	.070
	5/16	.313	7.950	.635	.191	.084	.088
	3/8	.375	9.525	.762	.230	.094	.106
	7/16	.438	11.125	.812	.223	.094	.103
	1/2	.500	12.700	.875	.223	.106	.103
	9/16	.563	14.300	1.000	.260	.118	.120
	5/8	.625	15.875	1.125	.298	.133	.137
	3/4	.750	19.050	1.375	.372	.149	.171

Fig. 3-23 Flat-head screw size chart

All sizes are maximum limit

ROUND-HEAD SCREWS

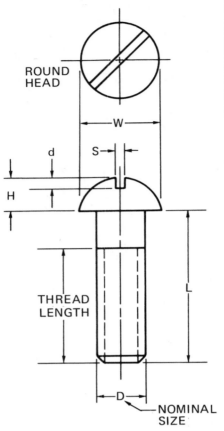

Fig. 3-24 Round-head screw representation

Type	Nom. Size (Inch)	— D —		— W —	— H —	— S —	— d —
		Inch	mm	Inch	Inch	Inch	Inch
Mach.	0	.060	1.524	.113	.053	.023	.039
	1	.073	1.854	.138	.061	.026	.044
	2	.086	2.184	.162	.069	.031	.048
	3	.099	2.514	.187	.078	.035	.053
	4	.112	2.845	.211	.086	.039	.058
	5	.125	3.175	.236	.095	.043	.063
	6	.138	3.505	.260	.103	.048	.068
	8	.164	4.166	.309	.120	.054	.077
	10	.190	4.826	.359	.137	.060	.087
	12	.216	5.486	.408	.153	.067	.096
Cap	1/4	.250	6.350	.472	.175	.075	.109
	5/16	.313	7.950	.590	.216	.084	.132
	3/8	.375	9.525	.708	.256	.094	.155
	7/16	.438	11.125	.750	.328	.094	.196
	1/2	.500	12.700	.813	.355	.106	.211
	9/16	.563	14.300	.938	.410	.118	.242
	5/8	.625	15.875	1.000	.438	.133	.258
	3/4	.750	19.050	1.250	.547	.149	.320

All sizes are maximum limit

Fig. 3-24 Round-head screw representation

FILLISTER-HEAD SCREWS

FILLISTER HEAD

Fig. 3-26 Fillister-head screw representation

Type	Nom. Size (Inch)	— D —		— W —	— H —	— h —	— S —	— d —
		Inch	mm	Inch	Inch	Inch	Inch	Inch
Mach.	0	.060	1.524	.096	.059	.045	.023	.025
	1	.073	1.854	.118	.071	.053	.026	.031
	2	.086	2.184	.140	.083	.062	.031	.037
	3	.099	2.514	.161	.095	.070	.035	.043
	4	.112	2.845	.183	.107	.079	.039	.048
	5	.125	3.175	.205	.120	.088	.043	.054
	6	.138	3.505	.226	.132	.096	.048	.060
	8	.164	4.166	.270	.156	.113	.054	.071
	10	.190	4.826	.313	.180	.130	.060	.083
	12	.216	5.486	.357	.205	.148	.067	.094
Cap	1/4	.250	6.350	.414	.237	.170	.075	.109
	5/16	.313	7.950	.518	.295	.211	.084	.137
	3/8	.375	9.525	.622	.355	.253	.094	.164
	7/16	.438	11.125	.625	.368	.265	.094	.170
	1/2	.500	12.700	.750	.412	.297	.106	.190
	9/16	.563	14.300	.812	.466	.336	.118	.214
	5/8	.625	15.875	.875	.521	.375	.133	.240
	3/4	.750	19.050	1.000	.612	.441	.149	.281

All sizes are maximum limit

Fig. 3-27 Fillister-head screw size chart

OVAL-HEAD SCREWS

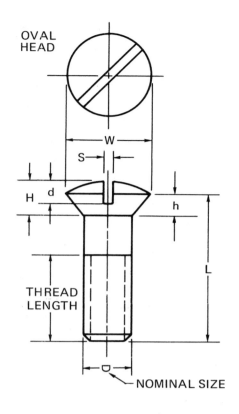

OVAL
HEAD

W

S

H d

h

L

THREAD
LENGTH

NOMINAL SIZE

Fig. 3-28 Oval-head screw representation

Type	Nom. Size (Inch)	– D –		– W –	– H –	– h –	– S –	– d –
		Inch	mm	Inch	Inch	Inch	Inch	Inch
Mach.	0	.060	1.524	.119	.056	.035	.023	.030
	1	.073	1.854	.146	.068	.043	.026	.038
	2	.086	2.184	.172	.080	.051	.031	.045
	3	.099	2.514	.199	.092	.059	.035	.052
	4	.112	2.845	.225	.104	.067	.039	.059
	5	.125	3.175	.252	.116	.075	.043	.067
	6	.138	3.505	.279	.128	.083	.048	.074
	8	.164	4.166	.332	.152	.100	.054	.088
	10	.190	4.826	.385	.176	.116	.060	.103
	12	.216	5.486	.438	.200	.132	.067	.117
Cap	1/4	.250	6.350	.507	.232	.153	.075	.136
	5/16	.313	7.950	.635	.290	.191	.084	.171
	3/8	.375	9.525	.762	.347	.230	.094	.206
	7/16	.438	11.125	.812	.345	.223	.094	.210
	1/2	.500	12.700	.875	.354	.223	.106	.216
	9/16	.563	14.300	1.000	.410	.260	.118	.250
	5/8	.625	15.875	1.125	.467	.298	.133	.285
	3/4	.750	19.050	1.375	.578	.372	.149	.353

All sizes are maximum limit

Fig. 3-29 Oval-head screw size chart

HEX-HEAD SCREWS AND BOLTS

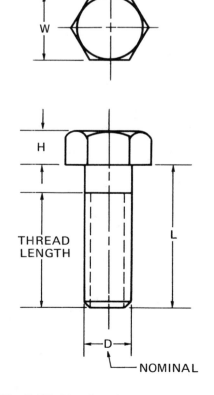

HEX HEAD

W

H

THREAD
LENGTH

L

D

NOMINAL

Fig. 3-30 Hex-head representation

Type	Nom. Size (Inch)	– D –		– W –		– H –	
		Inch	mm	Inch	mm	Inch	mm
Cap Screws	1/4	.260	6.60	.438	11.11	.163	4.14
	5/16	.324	8.23	.500	12.70	.211	5.36
	3/8	.388	9.86	.563	14.28	.243	6.17
	7/16	.452	11.48	.625	15.87	.291	7.39
	1/2	.515	13.08	.750	19.05	.323	8.20
	9/16	.577	14.66	.813	20.64	.371	9.42
	5/8	.642	16.31	.938	23.81	.403	10.24
	3/4	.768	19.51	1.125	28.57	.483	12.26
	7/8	.895	22.73	1.313	33.34	.563	14.30
	1	1.022	25.95	1.500	38.10	.627	15.93
	1 1/8	1.149	29.18	1.688	42.86	.718	18.23
	1 1/4	1.277	32.43	1.875	47.63	.813	20.65
Bolts	1 3/8	1.404	35.66	2.063	52.38	.940	23.88
	1 1/2	1.531	38.88	2.250	57.15	1.036	26.31
	1 3/4	1.785	45.34	2.625	66.68	1.196	30.37
	2	2.039	51.79	3.000	76.20	1.388	35.25
	2 1/4	2.305	58.54	3.375	85.72	1.548	39.32
	2 1/2	2.559	64.99	3.750	95.25	1.708	43.38
	2 3/4	2.827	71.80	4.125	104.78	1.869	47.47
	3	3.081	78.25	4.500	114.30	2.060	52.32

Fig. 3-31 Hex-head screw and bolt size chart

SQUARE-HEAD SCREWS AND BOLTS

SQUARE HEAD

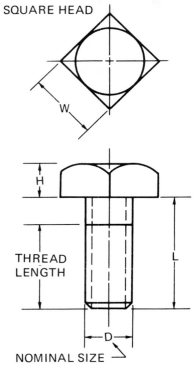

Fig. 3-32 Square-head representation

Type	Nom. Size (Inch)	– D –		– W –		– H –	
		Inch	mm	Inch	mm	Inch	mm
Cap Screws	1/4	.260	6.60	.375	9.52	.188	4.77
	5/16	.324	8.23	.500	12.70	.220	5.58
	3/8	.388	9.86	.563	14.28	.268	6.81
	7/16	.452	11.48	.625	15.87	.316	8.03
	1/2	.515	13.08	.750	19.05	.348	8.84
	5/8	.642	16.31	.938	23.81	.444	11.28
	3/4	.768	19.51	1.125	28.57	.524	13.31
	7/8	.895	22.73	1.313	33.34	.620	15.75
	1	1.022	25.95	1.500	38.10	.684	17.38
	1 1/8	1.149	29.18	1.688	42.86	.780	19.81
	1 1/4	1.277	32.43	1.875	47.63	.876	22.22
Bolts	1 3/8	1.404	35.66	2.063	52.38	.940	23.88
	1 1/2	1.531	38.88	2.250	57.15	1.036	26.31
	1 5/8	1.658	42.11	2.438	61.91	1.132	28.75

Fig. 3-33 Square-head screw and bolt size chart

SOCKET-HEAD SCREWS

SOCKET HEAD

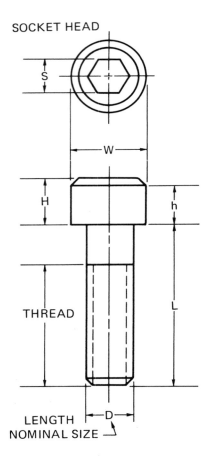

Fig. 3-34 Socket-head screw representation

Type	Nom. Size (Inch)	– D –		– W –	– H –	– h –	– S –
		Inch	mm	Inch	Inch	Inch	Inch
Mach.	0	.060	1.524	.096	.060	.055	.051
	1	.073	1.854	.118	.073	.067	.051
	2	.086	2.184	.140	.086	.079	.063
	3	.099	2.514	.161	.099	.091	.079
	4	.112	2.845	.183	.112	.103	.079
	5	.125	3.175	.205	.125	.115	.095
	6	.138	3.505	.226	.138	.127	.095
	8	.164	4.166	.270	.164	.150	.127
	10	.190	4.826	.313	.190	.174	.158
	12	.216	5.486	.344	.216	.198	.158
Cap	1/4	.250	6.350	.375	.250	.229	.189
	5/16	.313	7.950	.438	.313	.286	.220
	3/8	.375	9.525	.563	.375	.344	.315
	7/16	.438	11.125	.625	.438	.401	.315
	1/2	.500	12.700	.750	.500	.458	.378
	9/16	.563	14.300	.813	.563	.516	.378
	5/8	.625	15.875	.875	.625	.573	.503
	3/4	.750	19.050	1	.750	.688	.565

All sizes maximum limit

Fig. 3-35 Socket-head screw size chart

SETSCREWS

Setscrews are used to prevent rotary motion between mating parts, such as a pulley and shaft. They come in slotted, hex-socket, and square heads, figure 3-36. Note how all sizes are derived from the diameter (D).

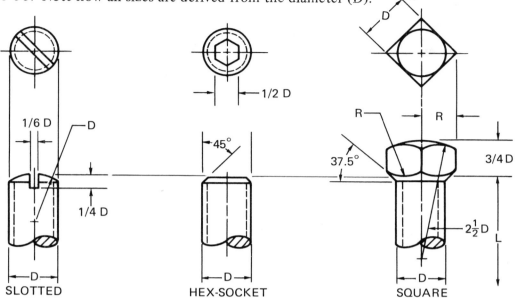

Fig. 3-36 Setscrew head types

SLOTTED   HEX-SOCKET   SQUARE

Points

Setscrew head styles come with a flat, cup, oval, cone, half-dog, or full-dog point, figure 31-37.

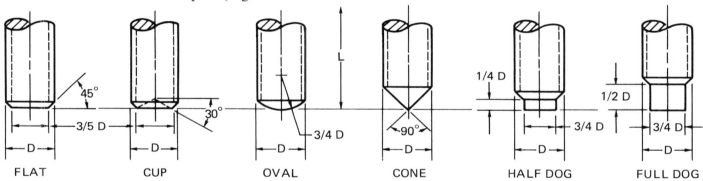

FLAT   CUP   OVAL   CONE   HALF DOG   FULL DOG

Fig. 3-37 Setscrew point types

Type	Nom. Size (Inch)	– D –		– UNC – (Coarse)	– UNF – (Fine)
		Inch	M.M.		
Mach.	0	.060	1.524	–	80
	1	.073	1.854	64	72
	2	.086	2.184	56	64
	3	.099	2.514	48	56
	4	.112	2.845	40	48
	5	.125	3.175	40	44
	6	.138	3.505	32	40
	8	.164	4.166	32	36
	10	.190	4.826	24	32
	12	.216	5.486	24	28
Cap	1/4	.250	6.350	20	28
	5/16	.313	7.950	18	24
	3/8	.375	9.525	16	24
	7/16	.438	11.125	14	20
	1/2	.500	12.700	13	20
	9/16	.563	14.300	12	18
	5/8	.625	15.875	11	18

Fig. 3-38 Setscrew size chart

HEXAGONAL AND SQUARE NUTS

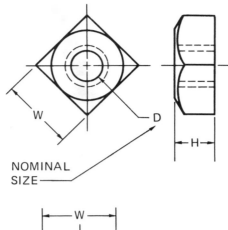

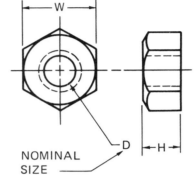

Fig. 3-39 Nut representation

Nom. Size (Inch)	– D –		– W –		– H –	
	Inch	mm	Inch	mm	Inch	mm
1/4	.260	6.60	.438	11.11	.235	5.96
5/16	.324	8.23	.563	14.28	.283	7.18
3/8	.388	9.86	.625	15.87	.346	8.78
7/16	.452	11.48	.750	19.05	.394	10.00
1/2	.515	13.08	.813	20.64	.458	11.63
9/16	.577	14.66	.875	22.23	.521	13.23
5/8	.642	16.31	1	25.40	.569	14.45
3/4	.768	19.51	1.125	28.57	.680	17.27
7/8	.895	22.73	1.313	33.34	.792	20.12
1	1.022	25.95	1.500	38.10	.903	22.94
1 1/8	1.149	29.18	1.688	42.86	1.030	26.16
1 1/4	1.277	32.43	1.875	47.63	1.126	28.60
1 3/8	1.404	35.66	2.063	52.38	1.237	31.42
1 1/2	1.531	38.88	2.250	57.15	1.348	34.24

Fig. 3-40 Nut size chart

PLAIN WASHERS

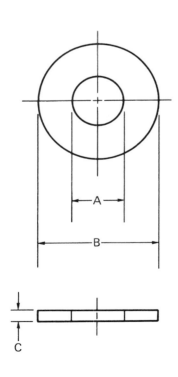

Fig. 3-41 Washer representation

Tol.	– A –		Tol.	– B –		Nom.	– C –	
	Inch	mm		Inch	mm		Inch	mm
+ OR – .005 (.127 M.M.)	3/32	2.38	+ OR – .010 (.254 M.M.)	1/4	6.35	NOMINAL SIZE	.020	.51
	1/8	3.18		5/16	7.94		.032	.81
	5/32	3.97		3/8	9.53		.049	1.24
	3/16	4.76		7/16	11.13		.049	1.24
	7/32	5.56		7/16	11.13		.049	1.24
+ OR – .010 (.254 M.M.)	1/4	6.35	+ OR – .010 (.254 M.M.)	1/2	12.70		.049	1.24
	9/32	7.14		5/8	15.88		.065	1.65
	5/16	7.94		7/8	22.23		.065	1.65
	11/32	8.73		11/16	17.47		.065	1.65
	3/8	9.53		3/4	19.05		.065	1.65
	13/32	10.32		13/16	20.64		.065	1.65
	7/16	11.11		1	25.40		.083	2.11
	1/2	12.70		1 1/4	31.75		.083	2.11
	17/32	13.50		1 1/16	26.99		.095	2.41
	9/16	14.28		1 3/8	34.93		.109	2.77
	5/8	15.88		1 3/8	34.93		.109	2.77
	11/16	17.46		1 1/2	38.10		.134	3.40
	13/16	20.64		1 1/2	38.10		.134	3.40
	15/16	23.81		2	50.80		.165	4.19
	1 1/16	26.98		2 1/4	57.15		.165	4.19

Fig. 3-42 Washer size chart

LOCK WASHERS

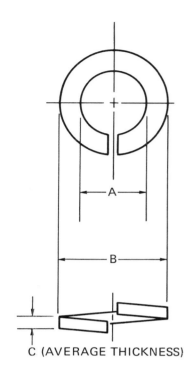

Fig. 3-43 Lock washer representation

— A —		— B —		— C —	
Inch	mm	Inch	mm	Inch	mm
.088	2.23	.175	4.44	.020	.51
.102	2.59	.198	5.02	.025	.64
.115	2.92	.212	5.38	.025	.64
.128	3.25	.239	6.07	.031	.79
.141	3.58	.253	6.43	.031	.79
.168	4.26	.296	7.52	.040	1.02
.194	4.92	.337	8.56	.047	1.02
.221	5.61	.380	9.65	.056	1.42
.255	6.47	.493	12.52	.062	1.57
.319	8.10	.591	15.01	.078	1.98
.382	9.70	.688	17.48	.094	2.38
.446	11.32	.784	19.91	.109	2.76
.509	12.92	.879	22.33	.125	3.18
.573	14.55	.979	24.87	.141	3.58
.636	16.15	1.086	27.58	.156	3.96
.700	17.78	1.184	30.07	.172	4.37
.763	14.38	1.279	32.48	.188	4.78
.827	21.01	1.377	34.97	.203	5.15
.890	22.60	1.474	37.44	.219	5.56
1.017	25.83	1.672	42.47	.250	6.35

Fig. 3-44 Lock washer size chart

COTTER PIN

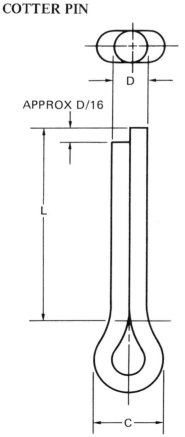

Fig. 3-45 Cotter pin representation

Nom. Size (Inch)	— D —		— d —		Hole Size	
	Inch	m.m.	Inch	m.m.	Inch	m.m.
1/32	.031	.787	.06	1.52	.047	1.193
3/64	.047	1.193	.09	2.28	.062	1.574
1/16	.062	1.574	.12	3.04	.078	2.006
5/64	.078	1.981	.16	4.06	.094	2.387
3/32	.094	2.387	.19	4.82	.109	2.768
7/64	.109	2.768	.22	5.58	.125	3.175
1/8	.125	3.175	.25	6.35	.141	3.581
9/64	.141	3.581	.28	7.11	.156	3.962
5/32	.156	3.962	.31	7.87	.172	4.368
3/16	.188	4.775	.38	9.65	.203	5.156
7/32	.219	5.562	.44	11.17	.234	5.943
1/4	.250	6.350	.50	12.70	.266	6.756
5/16	.312	7.925	.62	15.74	.312	7.925
3/8	.375	9.525	.75	19.05	.375	9.525
7/16	.438	11.125	.88	22.55	.438	11.125
1/2	.500	12.700	1.00	25.40	.500	12.700
5/8	.625	15.875	1.25	31.750	.625	15.875
3/4	.750	19.050	1.50	38.100	.750	19.050

Fig. 3-46 Cotter pin size chart

KEYS AND KEYSEATS

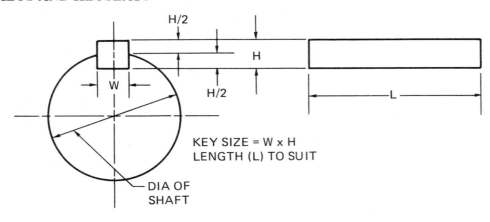

KEY SIZE = W x H
LENGTH (L) TO SUIT

DIA OF SHAFT

Fig. 3-47 Key representation

Shaft Nom. Size – DIA. –		Square (W = H)	Type	Square Key		Tolerance
From	To & Incl.			From	To & Incl.	
5/16 (8)	7/16 (11)	3/32 (2.38)	Bar Stock	—	3/4 (19.05)	+.000 – .002 (+.0000 – .0254)
7/16 (11)	9/16 (14)	1/8 (3.175)		3/4 (19.05)	1 1/2 (38.1)	+.000 – .003 (+.0000 – .0762)
9/16 (14)	7/8 (22)	3/16 (4.76)		1 1/2 (38.1)	2 1/2 (63.5)	+.000 – .004 (+.0000 – .1016)
7/8 (22)	1 1/4 (32)	1/4 (6.35)		2 1/2 (63.5)	3 1/2 (88.9)	+.000 – .006 (+.0000 – .1524)
1 1/4 (32)	1 3/8 (35)	5/16 (7.94)	Keystock	—	1 1/4 (31.75)	+.001 – .000 (+.0254 – .0000)
1 3/8 (35)	1 3/4 (44)	3/8 (9.53)		1 1/4 (31.75)	3 (76.2)	+.002 – .000 (+.0508 – .0000)
1 3/4 (44)	2 1/4 (57)	1/2 (12.7)		3 (76.2)	3 1/2 (88.9)	+.003 – .000 (+.0762 – .0000)
2 1/4 (57)	2 3/4 (70)	5/8 (15.88)				
2 3/4 (70)	3 1/4 (82)	3/4 (19.05)				
3 1/4 (82)	3 3/4 (95)	7/8 (22.23)				

(Figures in parenthesis = mm)

Fig. 3-48 Key size chart

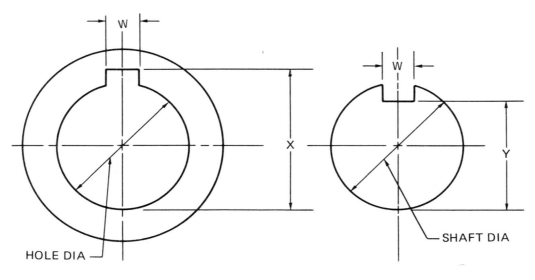

HOLE DIA

SHAFT DIA

Fig. 3-49 Keyway representation

Nom. Size	- DIA. - (Shaft)		'X' (Collar)		'Y' (Shaft)	
(Inch)	Inch	mm	Inch	mm	Inch	mm
1/2	.500	12.700	.560	14.224	.430	10.922
9/16	.562	14.290	.623	15.824	.493	12.522
5/8	.625	15.875	.709	18.008	.517	13.132
11/16	.688	17.470	.773	18.618	.581	14.757
3/4	.750	19.050	.837	21.259	.644	16.357
13/16	.812	20.640	.900	22.860	.708	17.983
7/8	.875	22.225	.964	24.485	.771	19.583
15/16	.938	23.820	1.051	26.695	.791	20.091
1	1.000	25.400	1.114	28.295	.859	21.818
1 1/16	1.062	26.985	1.178	29.921	.923	23.444
1 1/8	1.125	28.575	1.241	31.521	.986	25.044
1 3/16	1.188	30.165	1.304	33.121	1.049	26.644
1 1/4	1.250	31.750	1.367	34.722	1.112	28.244
1 5/16	1.312	33.340	1.455	36.957	1.137	28.879
1 3/8	1.375	34.923	1.518	38.557	1.201	30.505

Fig. 3-50 Keyway size chart

DRAWING FASTENERS

When drawing various fasteners, the drafter must use standard drawing practices.

In drawing a hexagonal nut or bolt head, always draw both the front view and the right-side view, figure 3-51, even though the right-side view is actually incorrect as shown. If drawn correctly, the hexagonal head actually looks like a square head.

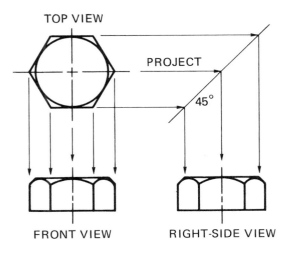

Fig. 3-51

Never draw a nut and bolt in section, figure 3-52, even if the cutting-plane line goes through the nut and bolt.

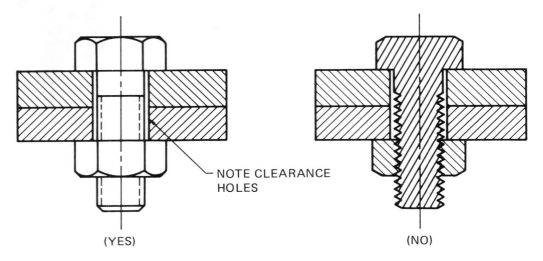

NOTE CLEARANCE HOLES

(YES)   (NO)

Fig. 3-52

A stud has two classes of fit. The section that screws into the part should have a tighter fit than the section at the nut end, figure 3-53. This is so the nut can be disassembled without removing the stud from the part. Sometimes fine threads are used at the stud end.

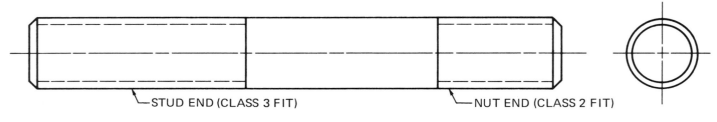

STUD END (CLASS 3 FIT)   NUT END (CLASS 2 FIT)

Fig. 3-53

When drawing various screw heads, draw the top view with the groove at a 45-degree angle as shown at B, figure 3-54, and the front view as if it were in position A. This is the standard method.

B

A

Fig. 3-54

The *standard thread depth* allowed in order to have the required strength is determined by the kind of material used. In steel, the depth *could* be thread diameter (minimum), but, in plastic or any other soft material, two times the thread diameter (minimum) must be used, figure 3-55. If possible, try to design it with more than minimum requirements, especially where safety is involved.

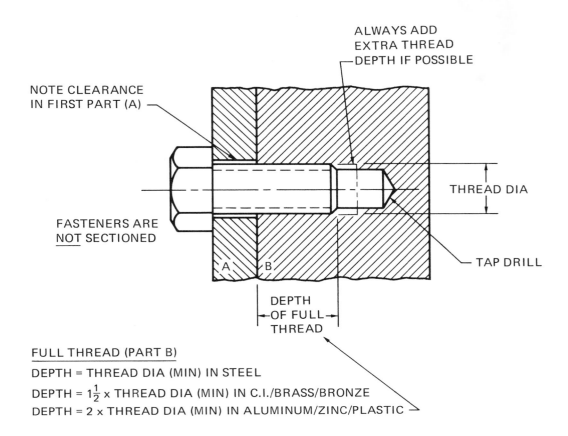

FULL THREAD (PART B)

DEPTH = THREAD DIA (MIN) IN STEEL

DEPTH = $1\frac{1}{2}$ x THREAD DIA (MIN) IN C.I./BRASS/BRONZE

DEPTH = 2 x THREAD DIA (MIN) IN ALUMINUM/ZINC/PLASTIC

Fig. 3-55

Any standard bolt or stud is usually designed around the thread lengths given in figure 3-56. The actual length is determined by the thread series (coarse, fine, or extra fine).

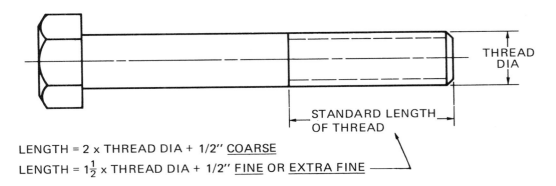

LENGTH = 2 x THREAD DIA + 1/2" <u>COARSE</u>

LENGTH = $1\frac{1}{2}$ x THREAD DIA + 1/2" <u>FINE</u> OR <u>EXTRA FINE</u>

Fig. 3-56

Thread Undercut or Relief

A simple spacer with threads looks like figure 3-57.

If an arm were attached and held in place with a nut, there probably will be a space between the shoulder and the arm, figure 3-58.

In order to eliminate this possibility, add an *undercut* or *thread relief,* figure 3-59.

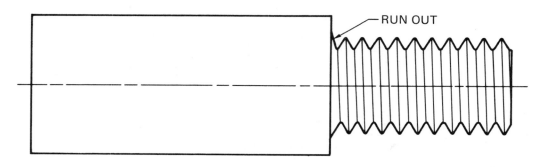

Fig. 3-57

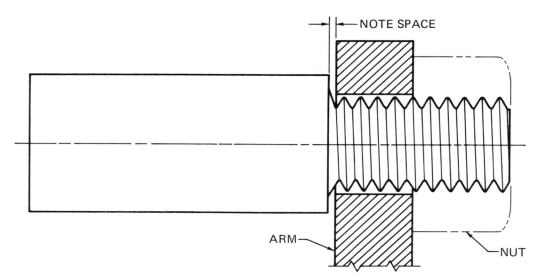

Fig. 3-58

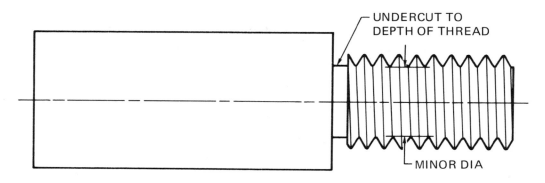

Fig. 3-59

Now the arm will fit against the shoulder tightly, figure 3-60. The thread relief is usually designed to the depth of the threads and called out as:

"XX WIDE X.XX DEEP THREAD RELIEF"

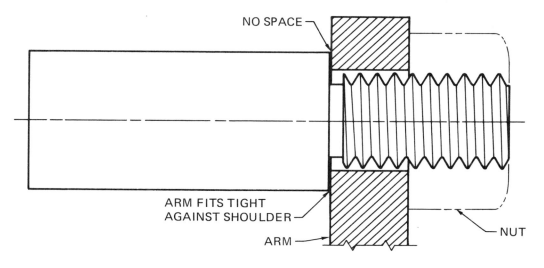

Fig. 3-60

Drawing a Fastener

The fastener callout and corresponding size charts provide the information needed to lay out a fastener. The drawing procedure given here uses a 1-UNC X 3 1/2 LG hexagonal bolt and nut as an example. For this fastener, refer to the chart for hex-head bolts on page 53 and the chart for hexagonal nuts on page 56.

Step 1. Lightly lay out the head diameter, length, and overall size of the nut.

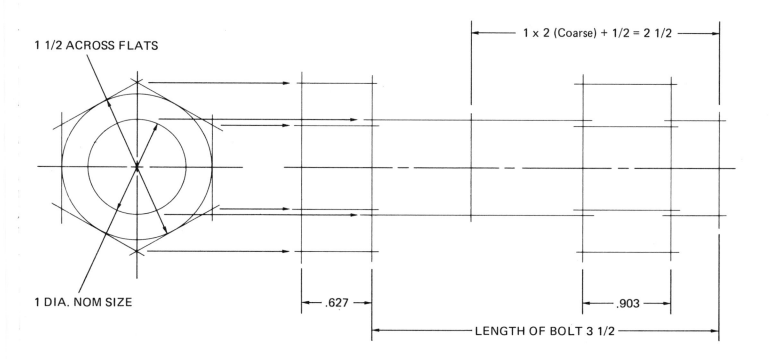

Step 2. Locate all radii, arcs, and chamfers as shown.

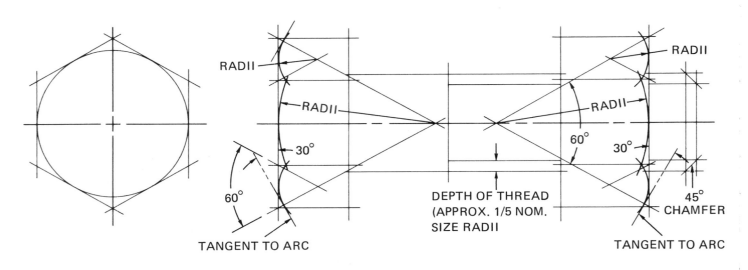

Step 3. Check all dimensions. Complete the drawing, using correct line weight.

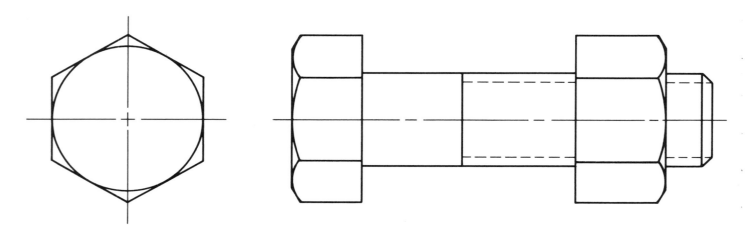

Review the charts on pages 51 through 59 once again. Note how each fastener is represented.

Practice Exercises 3-1 through 3-5

A drafter must be able to draw fasteners quickly and accurately using the various size charts. It is important to memorize the steps in drawing a fastener so time will not be lost reviewing each time one must be drawn. Using the size charts in figures 3-23 through 3-50, quickly and accurately draw the fasteners described in each problem. Use the center lines provided. In the guidelines below each exercise, carefully print the exact callout that is used in the actual drawing. Compare your work to the answers on page 67.

1. Draw a #10 flat-head screw that is 1 inch long with coarse, average-fitting threads.
2. Draw a 5/16-inch round-head screw with fine threads that is 2 1/4 inches long and a tight class of fit.
3. Draw a slotted-head screw that is 1 1/2 inches long, 3/4 — 10 thread, average class of fit.
4. Draw a plain washer for a 1/2-inch cap screw.
5. Draw a socket-head screw that is 3 inches long with tight-fitting fine threads, 1/2 inch in diameter.

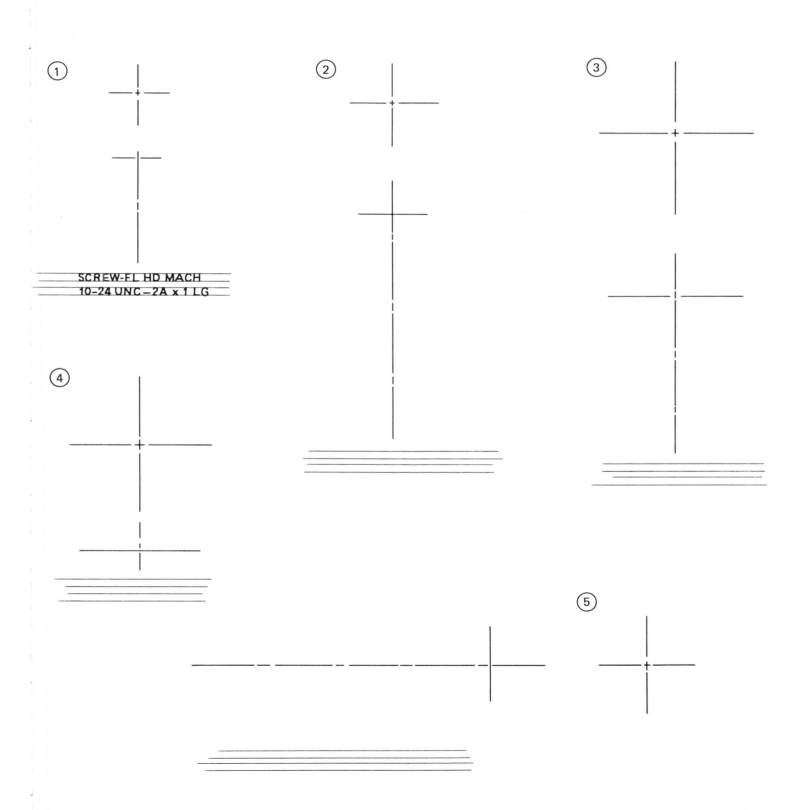

SCREW-FL HD MACH
10-24 UNC—2A x 1 LG

Practice Exercises 3-6, 3-7 and 3-8.

Lay out the following exercises in the spaces provided. Refer to the standard chart for nuts and bolts for dimensioning. Check line work. Compare your work to the answer on page 68.

Exercise 3-6

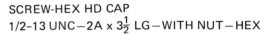

SCREW-HEX HD CAP
1/2-13 UNC—2A x $3\frac{1}{2}$ LG—WITH NUT—HEX

END OF BOLT

Exercise 3-7

SCREW-SQ HD CAP
5/8—11 UNC—2A x 3 LG WITH NUT — SQ

END OF
BOLT

Exercise 3-8

SCREW-SQ HD CAP
3/8—24 UNF—1A x $4\frac{1}{2}$ LG—WITH NUT — SQ

END OF
BOLT

ANSWERS TO PRACTICE EXERCISES

Exercises 3-1 through 3-5

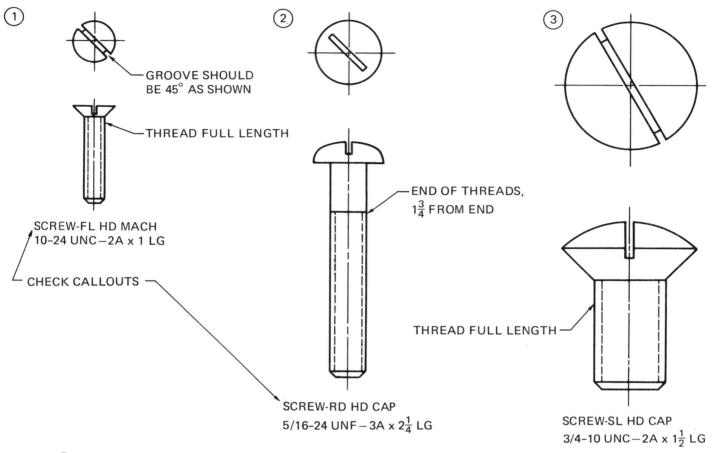

① GROOVE SHOULD
BE 45° AS SHOWN

THREAD FULL LENGTH

SCREW-FL HD MACH
10–24 UNC—2A x 1 LG

CHECK CALLOUTS

② END OF THREADS,
$1\frac{3}{4}$ FROM END

SCREW-RD HD CAP
5/16–24 UNF—3A x $2\frac{1}{4}$ LG

③ THREAD FULL LENGTH

SCREW-SL HD CAP
3/4–10 UNC—2A x $1\frac{1}{2}$ LG

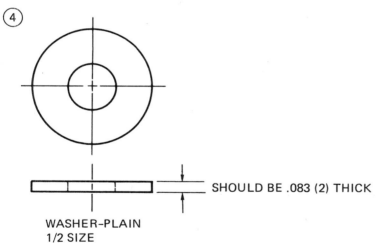

④ SHOULD BE .083 (2) THICK

WASHER–PLAIN
1/2 SIZE

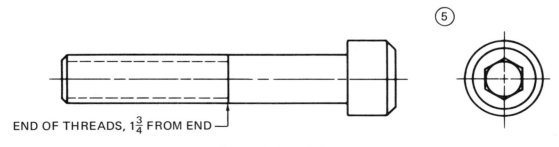

⑤ END OF THREADS, $1\frac{3}{4}$ FROM END

SCREW — SOC HD CAP
1/2 – 20 UNF – 3A x 3 LG

Exercises 3-6, 3-7, and 3-8

Exercise 3-6

SCREW-HEX HD CAP 1/2 — UNC — 2A x $3\frac{1}{2}$ LG   WITH NUT — HEX

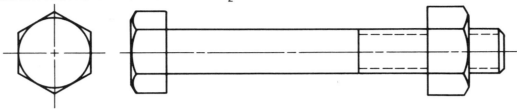

Exercise 3-7

SCREW-SQ HD CAP 5/8 — UNC — 2A x 3" LG WITH NUT SQ

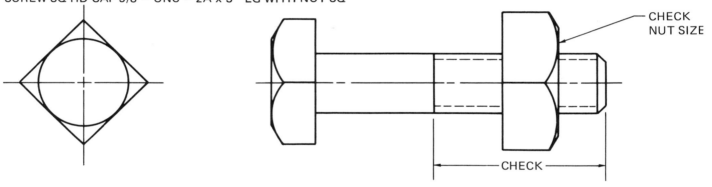

Exercise 3-8

SCREW-SQ HD CAP 3/8 — UNF — 1A x $4\frac{1}{2}$ LG WITH NUT — SQ

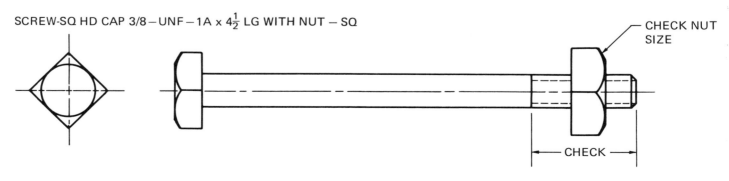

UNIT REVIEW

60-minute time limit

1. What does UNC and UNF mean?

2. Explain in full what 5/8 — 11 UNC — 2A × 2 1/2 LG means.

3. List the two major classifications of fasteners. Give examples of each.

4. A 2-inch (50.8) diameter shaft uses what size square key?

5. In addition to fastening parts together, what are two other uses of threads?

6. How is the depth of thread figured?

7. How many threads per inch are there in a 7/8 UNF thread?

8. What is the standard angle for ACME threads? For Unified National threads?

9. If you have a 1/4 — 20 UNC threaded screw and rotate it ten full turns, how far will the end travel?

10. If you have a 1/4 — 20 UNC double thread and rotate it ten full turns, how far will the end travel?

11. How is a screw called out that is 3 inches long, 3/8-inch diameter, coarse, right-hand threads, average fit, and round-head style?

12. What is the head diameter of a #12 fillister-head machine screw?

Before proceeding to the next unit:

_____ Instructor's approval

_____ Progress plotted

UNIT 4

PRECISION MEASUREMENT

OBJECTIVE

The student will learn to use and read micrometers and vernier calipers.

PRETEST

30-minute time limit

Identification

Identify the following parts and objects in the spaces provided.

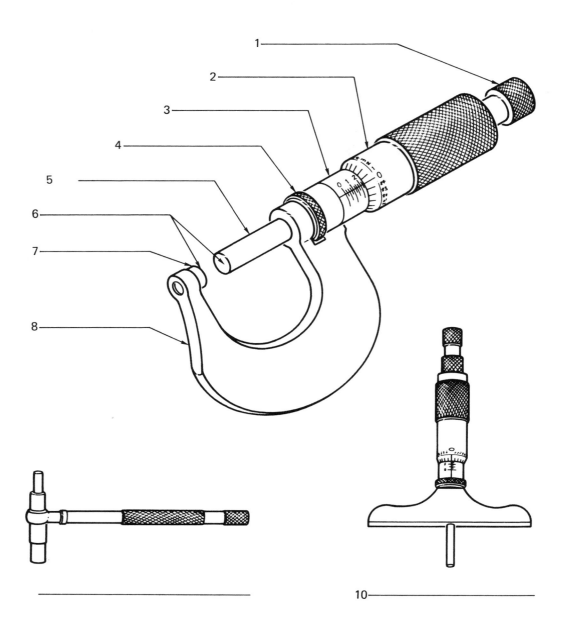

10————————————

11 HOW MANY MILLIMETRES IN AN INCH? —————————

Problems

In the spaces provided, determine the micrometer readings.

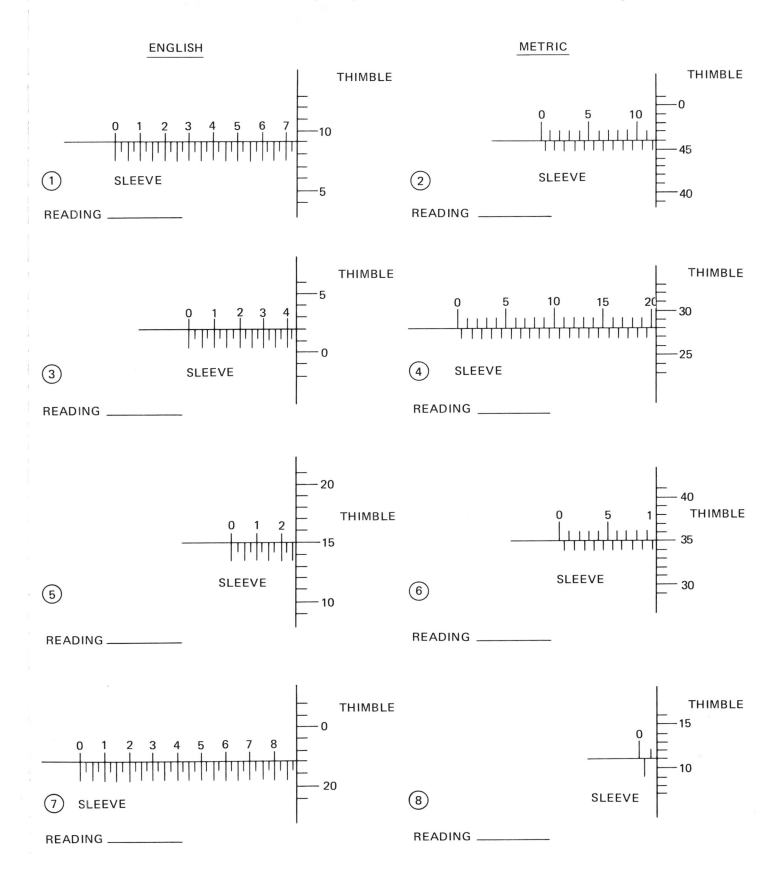

ENGLISH

METRIC

THIMBLE

THIMBLE

(1) SLEEVE

(2) SLEEVE

READING _____

READING _____

(3) SLEEVE

(4) SLEEVE

READING _____

READING _____

(5) SLEEVE

(6) SLEEVE

READING _____

READING _____

(7) SLEEVE

(8) SLEEVE

READING _____

READING _____

RELATED TERMS

Give a brief definition of each term as progress is made through the unit.

Pocket steel ruler _____

English system _____

Metric system _____

Metre _____

Kilometre _____

Millimetre _____

Calipers _____

Micrometer _____

Spindle _____

Sleeve _____

Thimble _____

Frame _____

Anvil _____

Telescoping gauges _____

Depth gauges _____

Vernier calipers _____

Vernier _____

Inside/outside measurements (calipers) _____

MEASURING

Unit 4 illustrates how to make precise measurements using both the inch (English system) and the millimetre (metric system). The metric system uses the *metre* (m) as its basic dimension. It is 3.281 feet long or about 3 3/8 inches longer than a yardstick. Its multiples, or parts, are expressed by adding prefixes. These prefixes represent equal steps of 1000 parts. The prefix for a thousand (1000) is *kilo;* the prefix for a thousandth (1/1000) is *milli.* One thousand metres (1000m), therefore, equals one kilometre (1.0 km). One thousandth of a metre (1/1000 m) equals one millimetre (1.0 mm). Comparing metric to English then:

One millimetre (1.0 mm) = 1/1000 metre - .03937 inch
One thousand millimetres (1000 mm) = 1.0 metre (1.0 m) = 3.281 feet
One thousand metres (1000 m) = 1.0 kilometre (1.0 km) = 3281.0 feet

POCKET STEEL RULER

The pocket steel ruler is the easiest of all measuring tools to use. The *inch scale,* figure 4-1, is six inches long and is graduated in 10ths and 100ths of an inch on one side and 32nds and 64ths on the other side.

The *metric scale* is 150 millimetres long (approximately six inches) and is graduated in millimetres and half millimetres on one side, figure 4-2. Sometimes metric pocket steel rulers are graduated in 64ths of an inch on the other side.

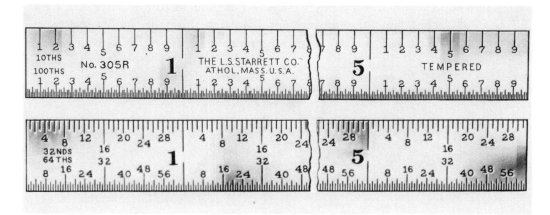

Fig. 4-1 Inch pocket ruler

Fig. 4-2 Metric pocket ruler

CALIPERS AND DIVIDERS

Figures 4-3 and 4-4 show two types of calipers used in a machine shop. They are spring-loaded with a nut to lock measurements into position. The *inside caliper,* figure 4-3, takes internal measurements, such as hole diameters and groove widths. The *outside caliper,* figure 4-4 takes external measurements, such as rod diameters and stock thickness measurements. Figure 4-5 shows a pair of dividers, a tool used to measure and transfer distances.

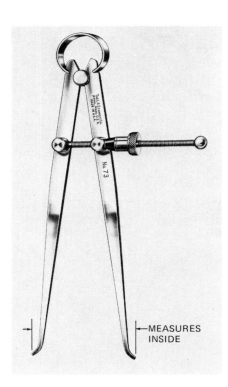

Fig. 4-3 Inside caliper

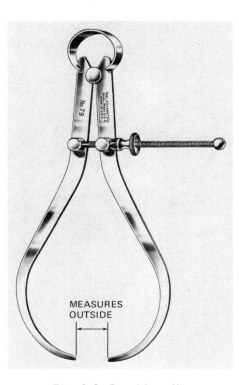

Fig. 4-4 Outside caliper

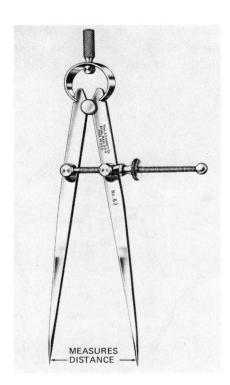

Fig. 4-5 Dividers

The calipers shown in figures 4-6 and 4-7 are similar to the calipers in figures 4-3 and 4-4 except that these are not spring-loaded. They are more sensitive and care must be taken not to alter the measurement once it is made with the caliper.

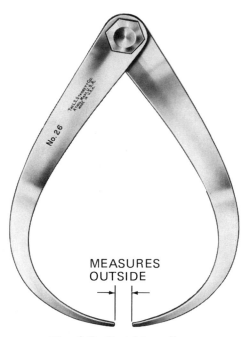

Fig. 4-6 Outside caliper

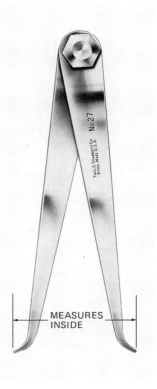

Fig. 4-7 Inside caliper

MICROMETER

The *micrometer* is a highly accurate screw which rotates inside a fixed nut, figure 4-8. As this screw is turned, it opens or closes a distance between the *measuring faces* of the anvil and spindle.

Place whatever is to be measured by the micrometer between the *anvil* and the *spindle*. Rotate the spindle by means of the *thimble* until the anvil and spindle come in contact with what is to be measured. The size is then determined by reading off the figures located on the *sleeve* and *thimble*.

Micrometers come in both English and metric graduations. They are manufactured with an English size range of 1 inch through 60 inches and a metric size range of 25 millimetres to 1500 millimetres. The micrometer is a very sensitive device and must be treated with extreme care.

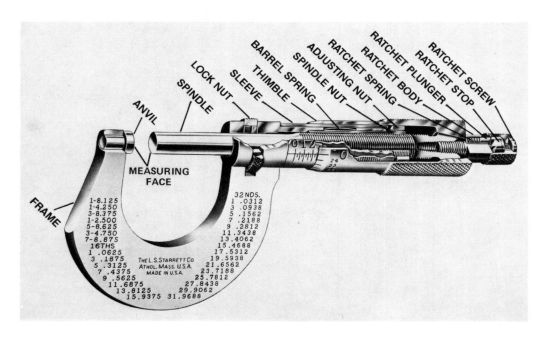

Fig. 4-8 Micrometer

Micrometers come in various shapes and sizes, figures 4-9 through 4-13.

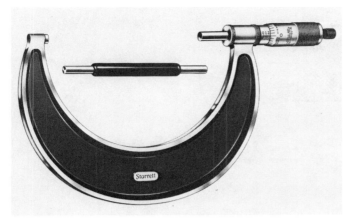

1″ TO 6″ MICROMETER
Fig. 4-9

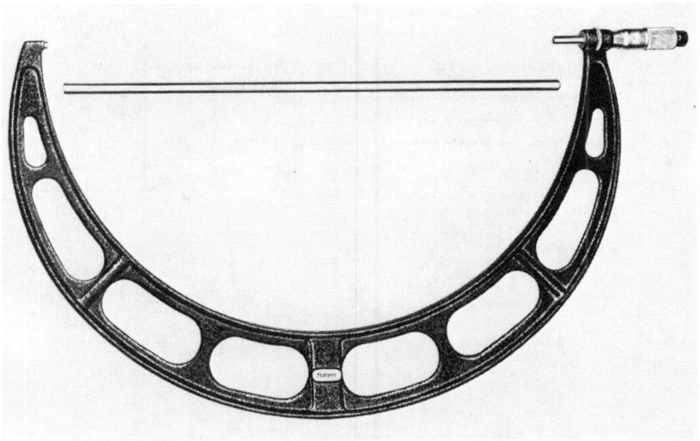

7″ TO 60″ MICROMETER
Fig. 4-10

Measurements of inside surfaces are made with a *telescoping gauge,* figure 4-12. The inside size is "locked," then the distance between ends is measured with a standard micrometer.

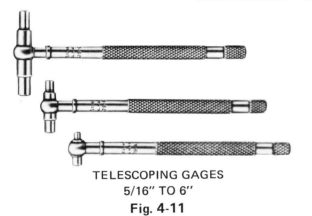

TELESCOPING GAGES
5/16″ TO 6″
Fig. 4-11

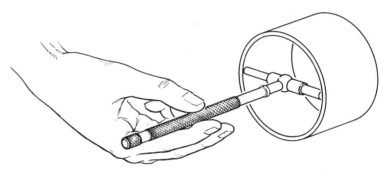

Fig. 4-12

The depth of holes, slots, and various projections are measured with a *micrometer depth gauge,* figure 4-13.

DEPTH GAGES
0″ TO 9″ SIZE

Fig. 4-13

How to Read a Micrometer — English System

The sleeve is divided into 40 even spaces. This corresponds to the pitch of the spindle which is 40 threads per inch. One complete turn of the thimble moves the spindle 1/40 or .025 inch, figure 4-4.

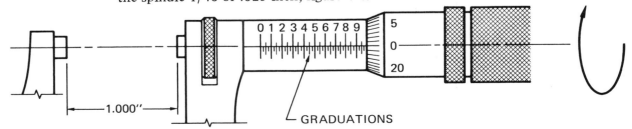

Fig. 4-14

1.000″

GRADUATIONS

The beveled edge of the thimble is divided into 25 equal parts. Each line equals .001 inch, figure 4-15.

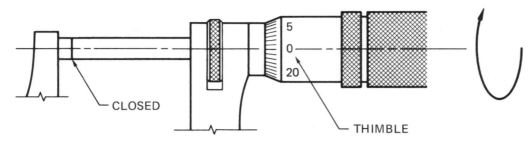

Fig. 4-15

To read the micrometer: Multiply the number of vertical lines on the sleeve by .025 inch. Add the number of thousandths indicated on the sleeve, figure 4-16.

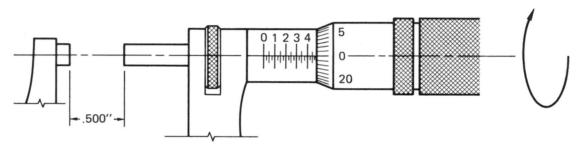

Fig. 4-16

How to Read a Micrometer — Metric System

The sleeve is divided into 50 even spaces. This corresponds to the pitch of the spindle which is 0.5 millimetre. One full turn of the thimble moves the spindle 0.5 millimetre, figure 4-17.

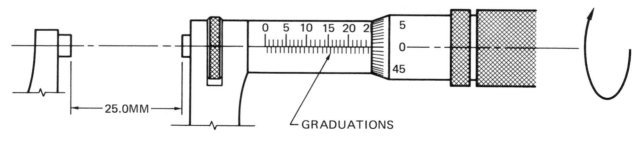

Fig. 4-17

The beveled edge of the thimble is divided into 50 equal parts. Each line equals 1/50 of 0.5 millimetre or 0.01 millimetre, figure 4-18.

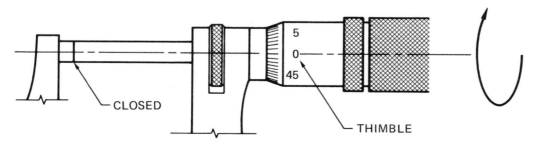

Fig. 4-18

To read the micrometer: Multiply the number of vertical lines on the sleeve by 0.5 millimetre. Add the number of millimetres indicated on the sleeve, figure 4-19.

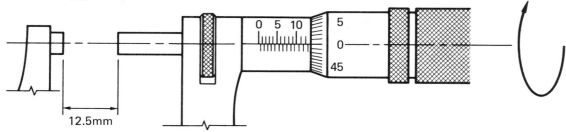

Fig. 4-19

SAMPLE READINGS

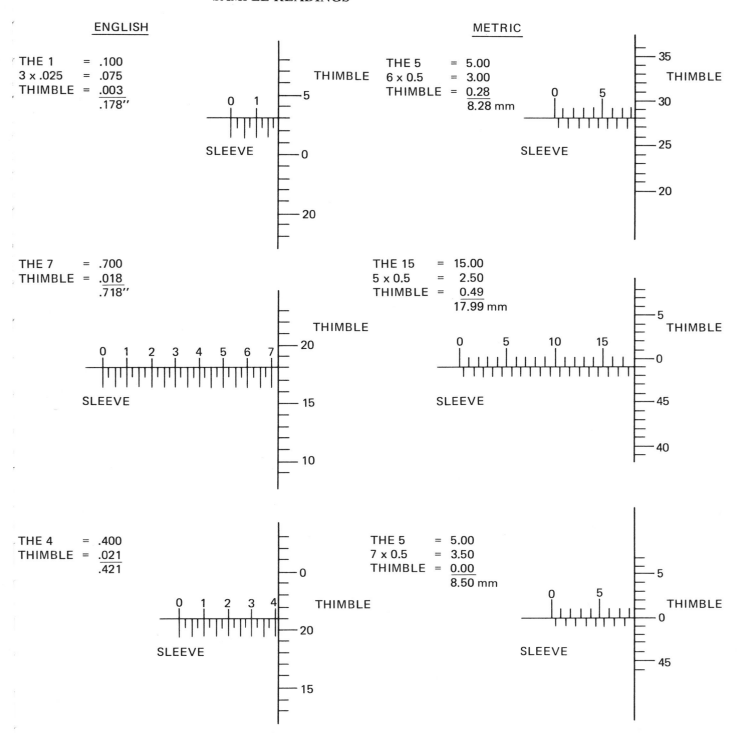

ENGLISH

THE 1 = .100
3 x .025 = .075
THIMBLE = .003
　　　　　　.178″

THE 7 = .700
THIMBLE = .018
　　　　　　.718″

THE 4 = .400
THIMBLE = .021
　　　　　　.421

METRIC

THE 5 = 5.00
6 x 0.5 = 3.00
THIMBLE = 0.28
　　　　　　8.28 mm

THE 15 = 15.00
5 x 0.5 = 2.50
THIMBLE = 0.49
　　　　　　17.99 mm

THE 5 = 5.00
7 x 0.5 = 3.50
THIMBLE = 0.00
　　　　　　8.50 mm

Practice Exercise 4-1

In the spaces provided, calculate what each reading is. Compare your work to the answers on page 83.

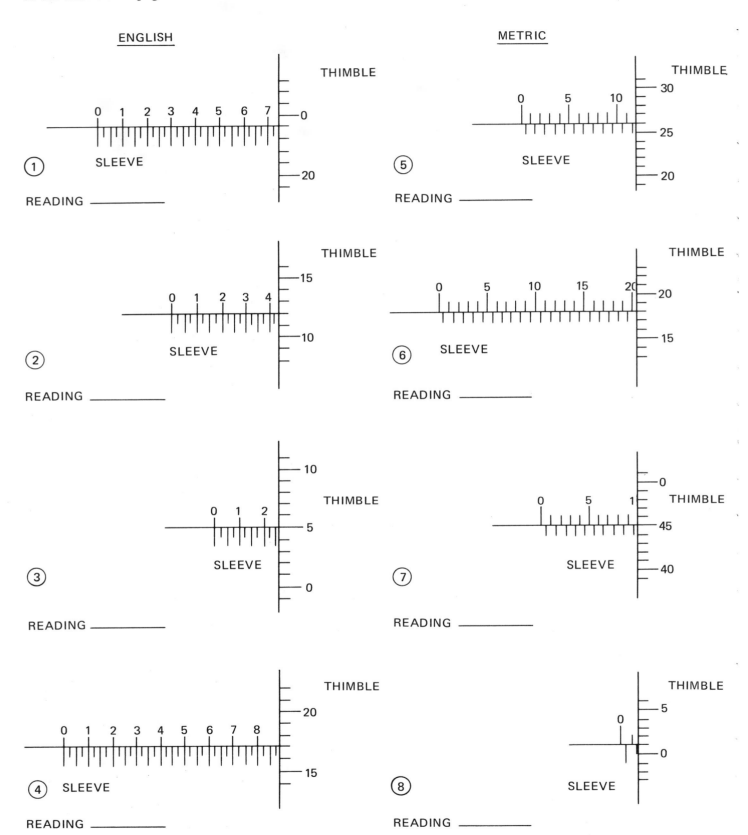

VERNIER CALIPERS

Vernier Calipers — English System

The vernier scale on the vernier caliper is read very much like a micrometer. The vernier, figure 4-20, reads to 1/1000 of an inch. The full inch is located on the main bar. Each inch is divided into four parts (.025 inch). The vernier scale is divided into 25 equal parts, each representing .001 inch.

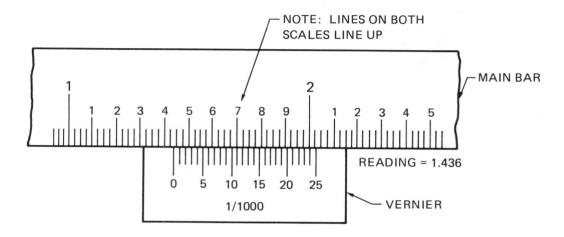

Fig. 4-20 Vernier caliper, English system

Using figure 4-20, a reading is calculated as follows:
From the main bar:

1. Record the full inches   — 1.000 (in thousandths)
2. Record the 1/10th inch — .400 (in thousandths)
3. Record the 1/25th inch — .025 (in thousandths)

From the vernier scale:

4. Count the number
   of graduations on
   the vernier scale
   from 0 to a line
   that coincides with a
   line on the main bar   — .011 (in thousandths)
                                  1.436 inches

Vernier Calipers — Metric System

The vernier scale in figure 4-21 is a metric venier and reads very much like the English vernier. It can be read to 0.02 millimetre. Each main bar graduation is in 0.5 millimetre with each 20th graduation numbered. The vernier is divided into 25 even parts, each representing 0.02 millimetre. To read the caliper, first record the total millimetres between zero on the main bar and zero on the vernier. Count the number of graduations on the vernier from zero to a line that coincides with a line on the main bar. Multiply that number times 0.02 millimetre and add this number to get the total reading.

Using the metric vernier scale in figure 4-21, the vernier zero is 41.5 millimetres past the location of the zero on the main bar. Therefore:

$$41.5 \text{ mm} + (9 \times 0.02 \text{ mm}) = 41.68 \text{ mm total reading.}$$

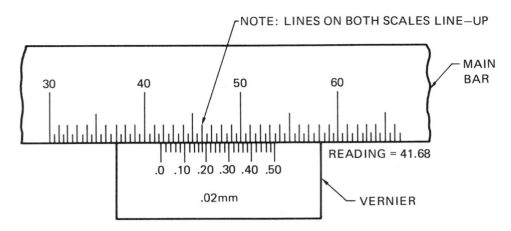

Fig. 4-21 Vernier caliper, metric system

Vernier calipers have the ability to measure both outside an object and inside an object, figures 4-22 and 4-23. When measuring an outside size, use the bottom scale. When measuring an inside size, use the top scale.

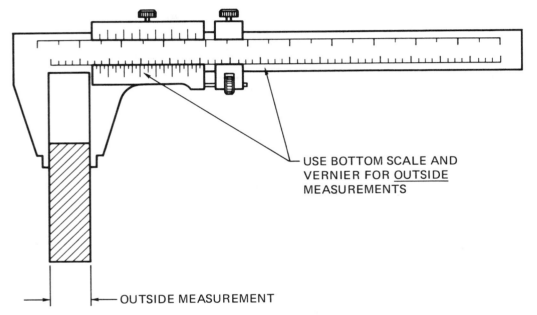

Fig. 4-22 Vernier caliper, outside measurement

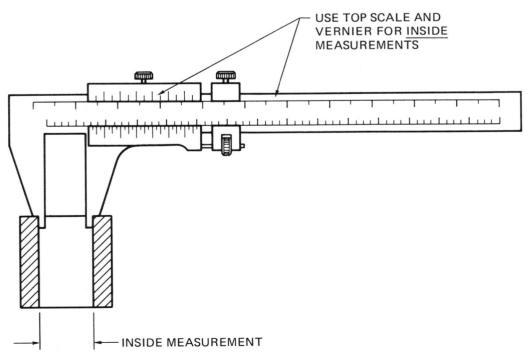

Fig. 4-23 Vernier caliper, inside measurement

ANSWERS TO PRACTICE EXERCISE 4-1

Answers should be 100 percent correct. After practicing with the micrometer, most calculations can be made directly from the instrument without doing the math as illustrated.

(1) Reading .749"

The "7"	=	.700
1 × .025	=	.025
Thimble	=	.024
		.749"

(2) Reading .437"

The "4"	=	.400
1 × .025	=	.025
Thimble	=	.012
		.437"

(3) Reading .255"

The "2"	=	.200
2 × .025	=	.050
Thimble	=	.005
		.255"

(4) Reading .892"

The "8"	=	.800
3 × .025	=	.075
Thimble	=	.017
		.892"

(5) Reading 11.76 mm

The "10"	=	10.00
3 × .5	=	1.50
Thimble	=	0.26
		11.76 mm

(6) Reading 20.18 mm

The "20"	=	20.00
0 × 0	=	0.00
Thimble	=	0.18
		20.18 mm

(7) Reading 9.95 mm

The "5"	=	5.00
9 × .5	=	4.50
Thimble	=	0.45
		9.95 mm

(8) Reading 1.51 mm

The "0"	=	0.00
3 × .5	=	1.50
Thimble	=	.01
		1.51 mm

UNIT REVIEW

15-minute time limit

1. What are the standard lengths for English and metric pocket steel rulers?

2. How long is a metre as compared to the English system of measurement?

3. In the metric system, what does the prefix *kilo* represent? The prefix *milli*?

4. What is a caliper?

5. On a micrometer, which two parts are used to determine a dimension?

6. Which size micrometer in the metric system is equal to the one-inch size in the English system?

7. How is the sleeve graduated in the metric and English systems?

8. How is the thimble graduated in the metric and English systems?

9. What is the degree of accuracy possible with a vernier English caliper? A vernier metric caliper?

10. How many inches are there in 254 millimetres? Round off the answers to the nearest inch.

Before proceeding to the next unit:

_____ Instructor's approval

_____ Progress plotted

UNIT 5

SPRINGS

OBJECTIVE

The student will learn to draw the major types of springs.

PRETEST

45-minute time limit

Construct a spring with the following specifications:

Free length 6.00″ (152)

Wire size .25″ (6)

10 active coils, 12 total coils

Plain open ends, not bent back

Solid length 3.00″ (76)

2.00″ (50) O.D.

1.50″ (38) I.D.

Left-hand windings

Compression-style spring

Material: Hard drawn steel — spring wire

Heat treatment: Heat to relieve coiling stresses

6.0 (152mm) APPROX FREE LENGTH

RELATED TERMS

Give a brief definition of each term as progress is made through the unit.

Compression spring _____

Extension spring _____

Torsion spring _____

Leaf spring _____

Spiral spring _____

Inside diameter (I.D.) _____

Outside diameter (O.D.) _____

Mean diameter _____

Wire size or diameter _____

Coils or turns _____

Total coils _____

Active coils _____

Free length _____

Loaded length _____

Solid length _____

Direction of windings _____

Open end _____

Closed end _____

Ground open end _____

Ground closed end _____

SPRINGS

A manufacturing company often uses a standard spring in assembling products. Occasionally, however, a spring must be designed by a drafter for a special job.

Figure 5-1 shows several types of springs.

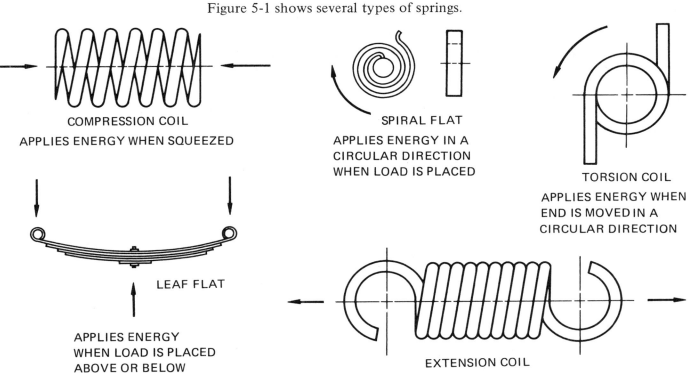

COMPRESSION COIL
APPLIES ENERGY WHEN SQUEEZED

SPIRAL FLAT
APPLIES ENERGY IN A
CIRCULAR DIRECTION
WHEN LOAD IS PLACED

TORSION COIL
APPLIES ENERGY WHEN
END IS MOVED IN A
CIRCULAR DIRECTION

LEAF FLAT
APPLIES ENERGY
WHEN LOAD IS PLACED
ABOVE OR BELOW

EXTENSION COIL
APPLIES ENERGY WHEN PULLED

Fig. 5-1 Types of springs

Study each part of the coil spring in figure 5-2.

Turn or *coil* is one complete turn about the center axis.

Total coils are the total number of turns or coils, starting from one point to the exact point on the next coil.

Active coils, in a compression spring, are usually the total coils minus the two end coils. Figure 5-2 has four active coils.

Free length is the overall length of the spring with no load on it. On a compression spring it is measured as noted in figure 5-2. On an extension spring, the measurement is taken from the inside of the hooks or ends.

Loaded length is the overall length of the spring with a given load applied to it.

Solid length, on a compression spring, is the length of the spring when it is completely compressed with each coil closed upon the next. In figure 5-2, if the wire diameter is .500 inch (12.7), the overall solid length is 3 inches (76.2).

Wire diameter is the diameter of the wire used to make the spring.

Outside diameter (O.D.) is the diameter of the outside of the coil spring.

Inside diameter (I.D.) is the diameter of the inside of the coil spring.

Mean diameter is the theoretical diameter of the spring measured to the center of the wire diameter. This diameter is used to draw the spring. To find the mean diameter, subtract the wire diameter from the outside diameter.

Direction of a spring describes whether the spring is wound left-hand or right-hand. Figure 5-2 is a left-hand spring.

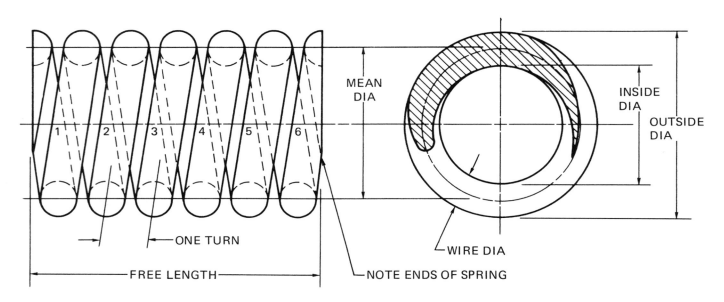

Fig. 5-2 Coil spring, detail drawing

Winding Direction

Springs can be wound left-hand or right-hand, figure 5-3. If the coil winding direction is not called out on a drawing, it will be manufactured with a right-hand winding.

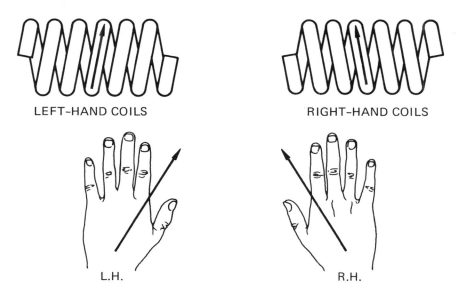

Fig. 5-3 Determining direction of winding

COMPRESSION SPRING ENDS

Compression springs have either open ends, closed ends, ground open ends, or ground closed ends, figure 5-4. Note that the ground springs are made from plain open or closed springs. The plain open-ended spring is the most economical to manufacture, though its use is limited. Springs with ground closed ends are the most stable and can be ground back for even more support.

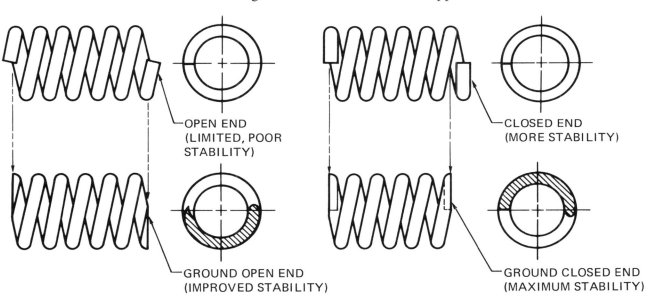

Fig. 5-4 Compression spring ends

DRAWING A COMPRESSION SPRING

The following drawing procedure can be used to lay out a compression spring. As an example, a spring with these specifications is used:

Overall length 4″	L.H. winding
8 total coils (6 active)	1 1/2″ O.C.
Plain open ends	1″ I.D.
Wire size .25″	

Step 1. Lightly lay out the overall length and mean diameter of the spring. Divide the length into even spaces. The number of even spaces will depend on the total number of coils and the type of end required:

For plain open ends:	2 × total coils + 1
For plain closed ends:	2 × total coils − 1
For ground open ends:	2 × total coils − 1
For ground closed ends:	2 × total coils − 1

The spring in this example therefore has 17 even spaces (2 × 8 = 16 + 1 = 17).

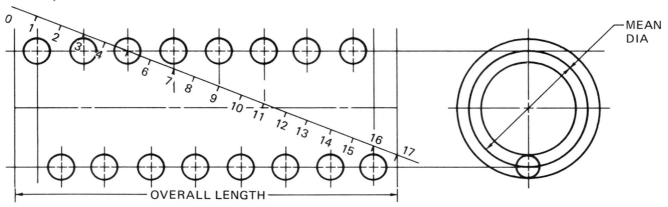

Step 2. Lightly draw circles at the top and bottom of each space to represent a cross section of each coil.

Step 3. Lightly draw the coil winding direction.

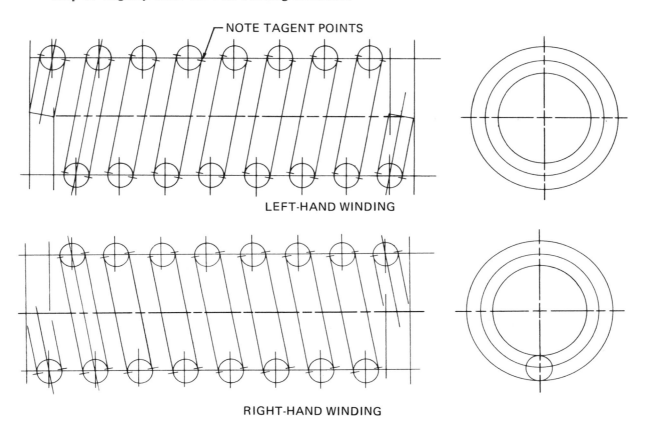

Step 4. Draw the style of end required. A plain open end is shown here. See figure 5-4 for other end styles.

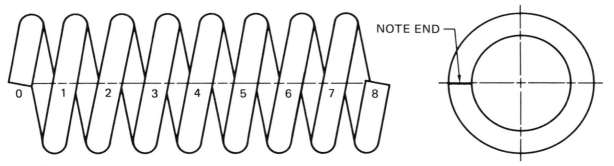

Step 5. Using correct line weight, finish the drawing. Add all dimensions and required notations.

Practice Exercises 5-1 through 5-4

After completing each exercise, compare your work to the answers on pages 102 through 105.

Exercise 5-1. Construct a plain open-ended spring with the following specifications:

Free length 7.00″ (178)

Wire size .25″ (6)

12 active coils, 14 total coils

Plain open ends, not bent back

Compression style spring

Material: Hard drawn steel spring wire

Heat treatment: Heat to relieve coiling stresses

Solid length 3.50″ (89)

2.00″ (50) O.D.

1.50″ (38) I.D.

R.H. windings

7.0 (178) APPROX FREE LENGTH

Exercise 5-2. Construct a ground open ended spring with the following specifications:

Free length 5.25" (134)

Ground open ends

L.H. winding

Wire size .375" (10)

2.75" (70) O.D.

Compression-style spring

4 active coils, 6 total coils

2.00" (50) I.D.

Material – hard drawn steel-spring wire

Heat treatment – heat to relieve coiling stresses

5.25 (134) APPROX FREE LENGTH

Exercise 5-3. Construct a spring with plain closed ends to the following specifications:

Free length 6.25" (159)
Wire size .375" (10)
7 active coils, 9 total coils
Plain closed ends
2.00" (50) O.D.
1.25" (32) I.D.
L.H. winding
Compression-style spring

6.25 (159) APPROX FREE LENGTH

Exercise 5-4. Construct a spring with ground closed ends to the following specifications:

Free length 3.5'' (89)
Wire size .438'' (11) diameter
2 active coils, 4 total coils
Ground closed ends
2.813'' (72) I.D.
3.688'' (94) O.D.
R.H. winding
Compression-style spring

3.5 (89mm) APPROX FREE LENGTH

EXTENSION SPRING ENDS

Figure 5-5 illustrates a few of the many types of ends used on *extension springs*. A loop completes a turn on itself while a hook is open. An extension spring has tight windings and is usually wound with right-hand windings.

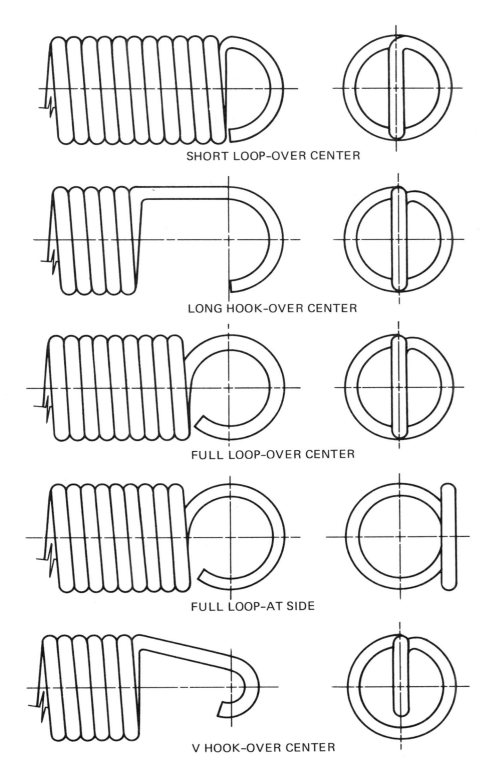

SHORT LOOP–OVER CENTER

LONG HOOK–OVER CENTER

FULL LOOP–OVER CENTER

FULL LOOP–AT SIDE

V HOOK–OVER CENTER

Fig. 5-5 Types of ends on extension springs

DRAWING AN EXTENSION SPRING

On a separate sheet of paper, draw a full loop over center extension spring following the steps given. Specifications:

Overall length approx. 4.75" (120 mm)
Full loop, over center
Wire size .188 (5)
R.H. winding (standard)
1.625" (41) O.D.
1.25" (32) I.D.
Extension spring style

Step 1. Make a rough sketch.

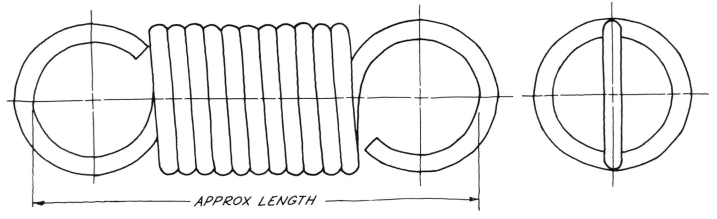

Step 2. Draw the end view (O.D./I.D.) and each end loop at the required length.

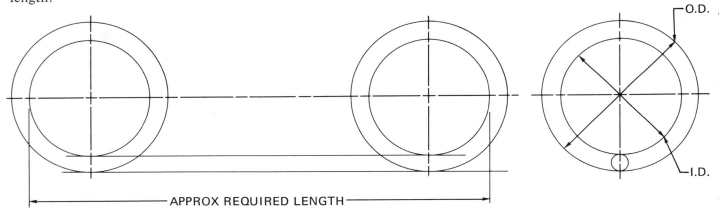

Step 3. Very carefully measure off the wire size springs and, with a circle template, lightly draw the approximate number of coils. Starting at the top from the right end, project to the left 1/2 wire size.

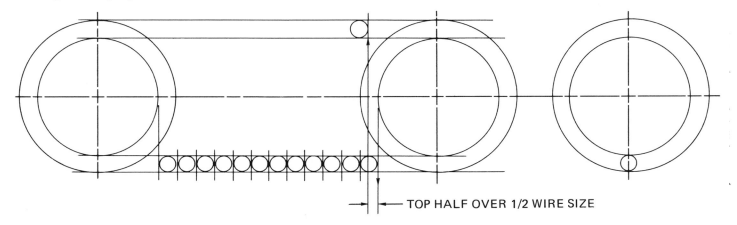

Step 4. Line up the drafting machine or triangle on the edge of the wire diameters (at X and Y) and lock on this angle. Lightly project up from wire diameters.

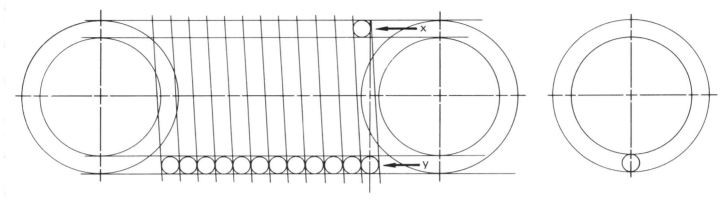

Step 5. Draw the wire diameters lightly. From the last wire size (A), project straight down. Adjust the end loop in. Note: Study the right end and be sure you understand how it is drawn (numbers 1, 2, 3, and 4).

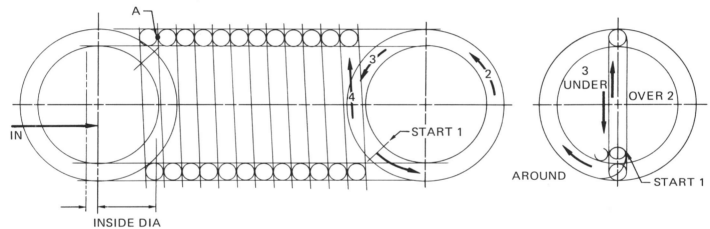

Step 6. Finish the drawing using correct line weight. Add dimensions and required notes.

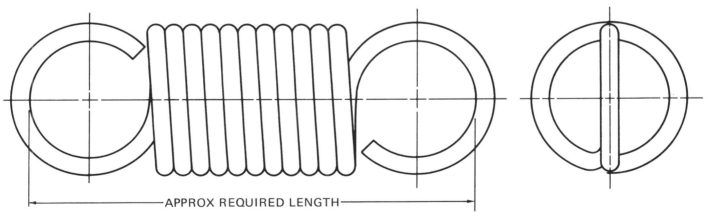

Practice Exercises 5-5 through 5-7

After completing each exercise, compare your work to the answers on pages 106 through 108.

Exercise 5-5. Construct an extension spring with the following specifications:

Approx. free length 7.00" (178)

Wire size .25" (6)

2.00" (50) O.D.

1.50" (38) I.D.

Full loop (both ends over center)

R.H. winding (standard)

Extension spring style

Material: Hard drawn steel spring wire

Heat treatment: Heat to relieve coiling stresses

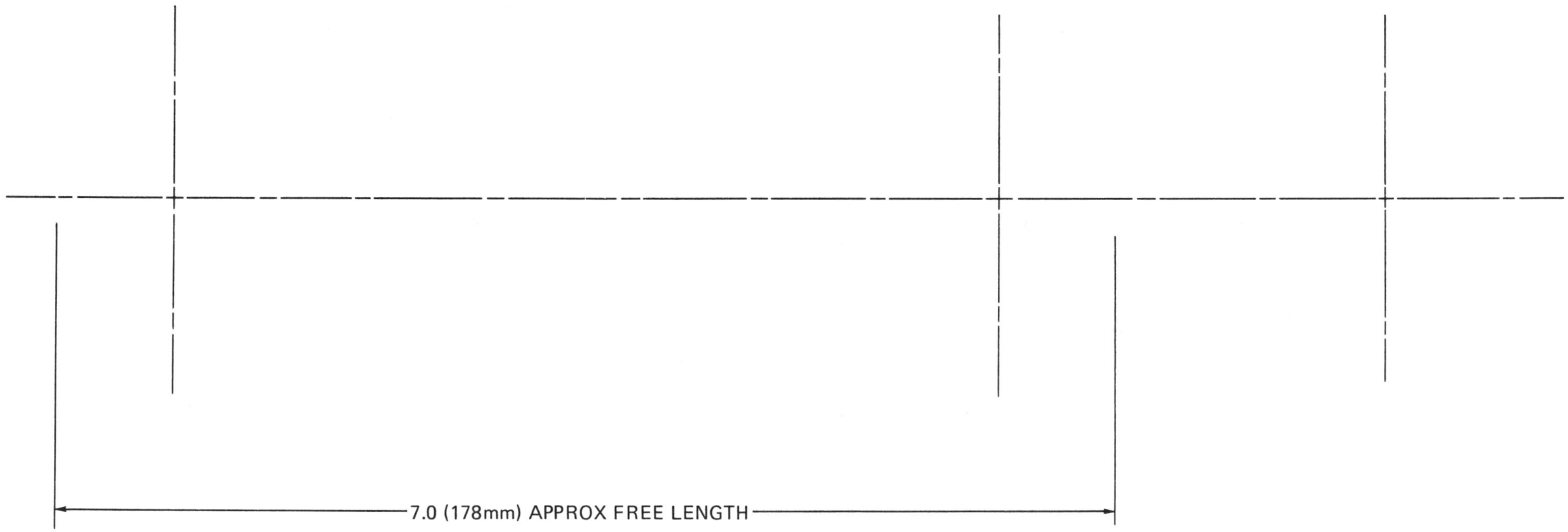

7.0 (178mm) APPROX FREE LENGTH

Exercise 5-6. Construct a spring with the following specifications:

Free length 5.875'' (149)

Wire size .188'' (5)

13 total coils, 2.625'' (67) coil length

1.625'' (41) O.C.

1.25'' (32) I.D.

Standard R.H. winding

Full loop, over center (right end)

Long hook, over center (left end)

Extension style spring

Material — hard drawn steel spring wire

Heat treatment — heat to relieve coiling stresses

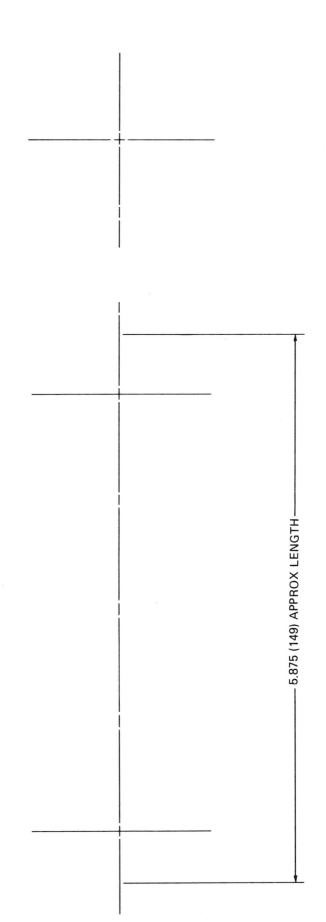

5.875 (149) APPROX LENGTH

Exercise 5-7. Construct a spring with the following specifications:

Free length 3.625" (92)   1.50" (38) I.D.

Wire size .25" (6)   R.H. winding

13 total coils   Torsion style spring per sketch

2" (50) O.D.

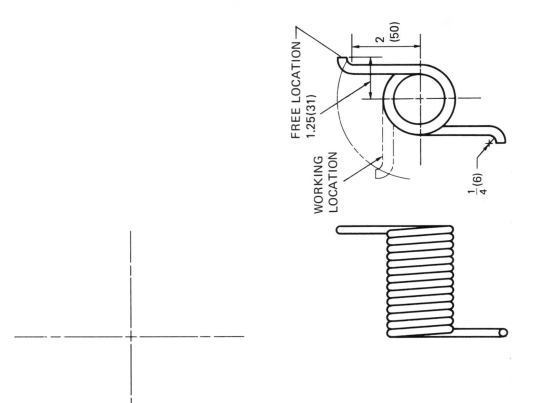

FREE LOCATION
1.25(31)

WORKING
LOCATION

2 (50)

$\frac{1}{4}$ (6)

STANDARD DRAFTING PROCEDURES

Some companies will not pay a drafter to draw a conventional representation of a spring, figure 5-6, because of the time and cost involved.

A short cut method to draw a spring is shown in figure 5-7. This represents the spring in figure 5-6, but takes less time to draw. Even in semiconventional spring representations (schematics) all dimensions and notes must be added.

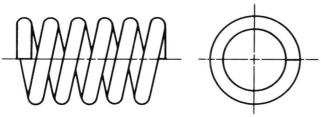

Fig. 5-6 Conventional spring drawing

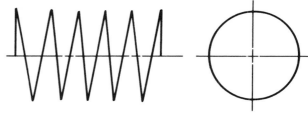

Fig. 5-7 Schematic spring drawing

Another method of drawing long springs rapidly is shown in figure 5-8.

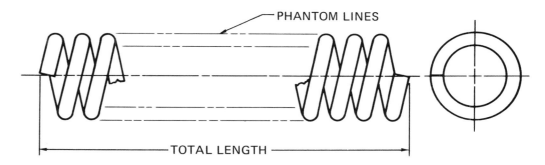

Fig. 5-8 Drawing long springs rapidly

Figure 5-9 shows a section view of a small spring (top) and a large spring (bottom). Both are right-hand springs but, since the back half of the springs are shown, they appear left-hand in section.

SECTION VIEW OF A SMALL SPRING

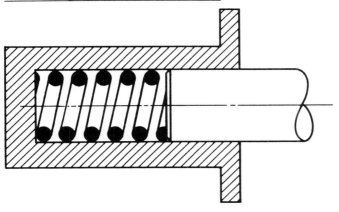

SECTION VIEW OF A LARGE SPRING

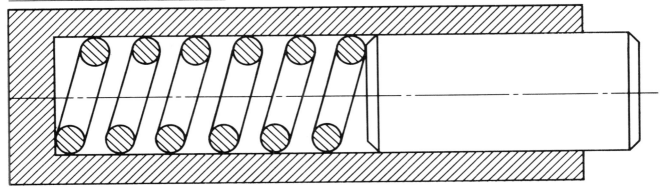

Fig. 5-9 Sectioning small and large springs

ANSWERS TO PRACTICE EXERCISES

Carefully compare your work to the answer for each exercise. Refer any questions to the instructor. Illustrate all dimensions and notes. Check line weight, neatness, and overall appearance.

Exercise 5-1

There are fourteen required turns (coils). The ends of the spring are plain open ends.

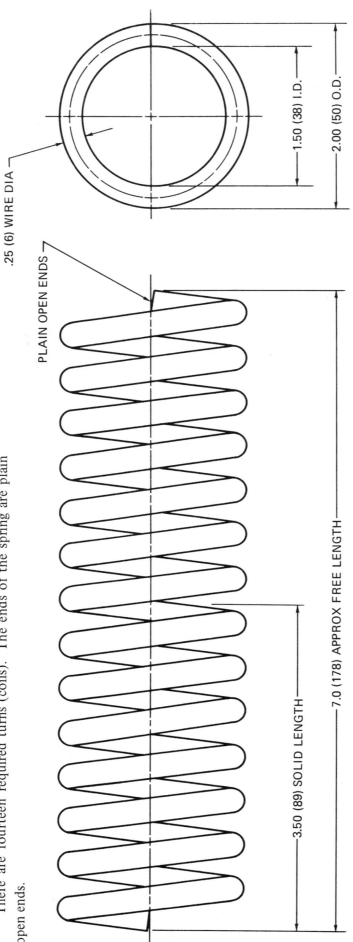

.25 (6) WIRE DIA

1.50 (38) I.D.

2.00 (50) O.D.

PLAIN OPEN ENDS

3.50 (89) SOLID LENGTH

7.0 (178) APPROX FREE LENGTH

MATERIAL: HARD DRAWN STEEL—SPRING WIRE HEAT TREAT: HEAT TO RELIEVE COILING STRESSES

Exercise 5-2

There are six required turns. The ends of the spring are ground open ends.

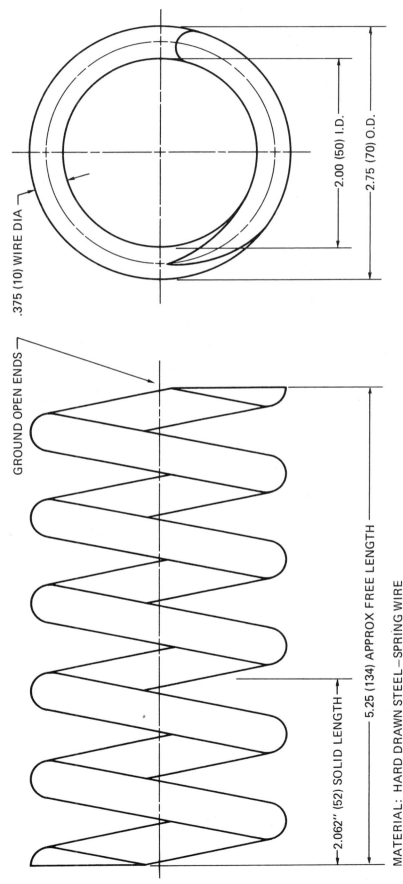

.375 (10) WIRE DIA

2.00 (50) I.D.

2.75 (70) O.D.

GROUND OPEN ENDS

5.25 (134) APPROX FREE LENGTH

2.062" (52) SOLID LENGTH

MATERIAL: HARD DRAWN STEEL —SPRING WIRE
HEAT TREAT: HEAT TO RELIEVE COILS STRESSES

Exercise 5-3

There are nine required turns. The ends of the spring are plain closed ends.

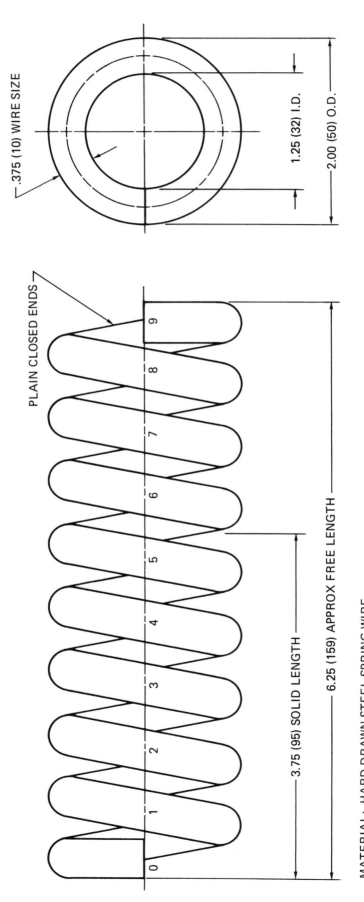

.375 (10) WIRE SIZE

PLAIN CLOSED ENDS

1.25 (32) I.D.

2.00 (50) O.D.

3.75 (95) SOLID LENGTH

6.25 (159) APPROX FREE LENGTH

MATERIAL: HARD DRAWN STEEL SPRING WIRE
HEAT TREAT: HEAT TO RELIEVE COILING STRESSES

Exercise 5-4

There are four required turns. The ends of the spring are closed ground ends.

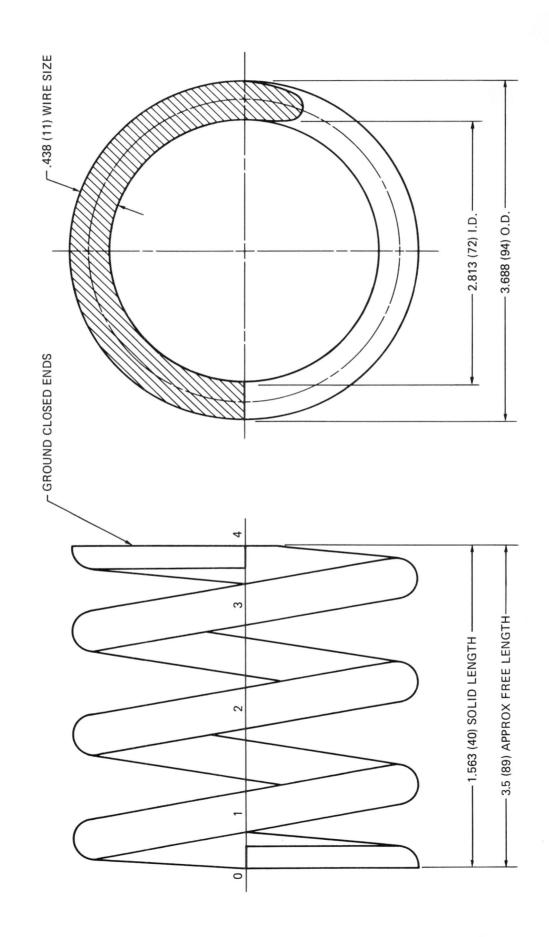

.438 (11) WIRE SIZE

2.813 (72) I.D.

3.688 (94) O.D.

GROUND CLOSED ENDS

1.563 (40) SOLID LENGTH

3.5 (89) APPROX FREE LENGTH

MATERIAL: HARD DRAWN STEEL SPRING WIRE
HEAT TREAT: HEAT TO RELIEVE COILING STRESS

Exercise 5-5

Include all required turns for this approximate free length. Check the ends of the spring for full loop over the center.

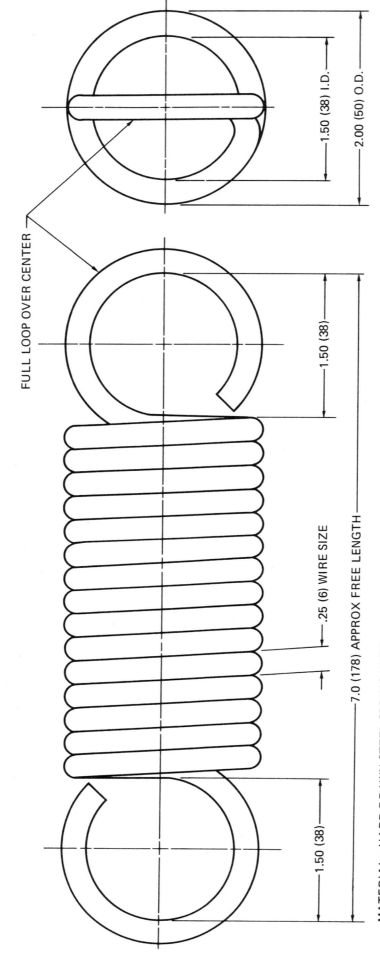

FULL LOOP OVER CENTER

1.50 (38) I.D.

2.00 (50) O.D.

1.50 (38)

.25 (6) WIRE SIZE

7.0 (178) APPROX FREE LENGTH

1.50 (38)

MATERIAL: HARD DRAWN STEEL SPRING WIRE
HEAT TREAT: HEAT TO RELEIVE COILING STRESSES

Exercise 5-6

Include all required turns (2.625 inch or 67 mm). Check the spring ends.

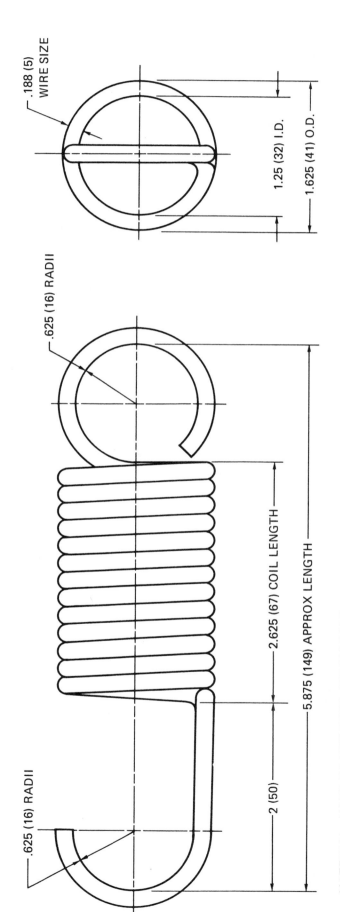

.188 (5) WIRE SIZE

1.25 (32) I.D.

1.625 (41) O.D.

.625 (16) RADII

2.625 (67) COIL LENGTH

5.875 (149) APPROX LENGTH

2 (50)

.625 (16) RADII

MATERIAL: HARD DRAWN STEEL SPRING WIRE
HEAT TREAT: HEAT TO RELIEVE COILING STRESSES

Exercise 5-7

Check required coils, ends of springs, notes, dimensions, and line weight.

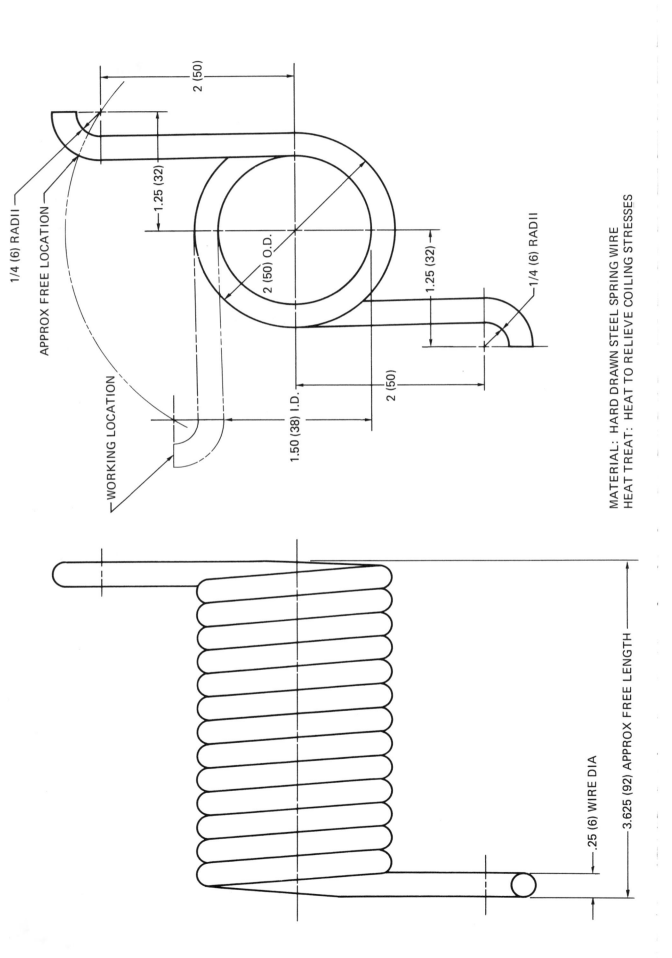

MATERIAL: HARD DRAWN STEEL SPRING WIRE
HEAT TREAT: HEAT TO RELIEVE COILING STRESSES

UNIT REVIEW

30-minute time limit

1. Which direction of winding is used most often on an extension spring?

2. What is the first thing to do when drawing a spring?

3. List four kinds of ends used on an extension spring.

4. How is a small spring drawn in section?

5. What must be included on spring drawings besides the dimensions and material?

6. List four kinds of ends used on a compression spring.

7. Explain solid length as it relates to a compression spring.

8. What is meant by free length?

9. Explain active coils.

10. What is the mean diameter?

Before proceeding to the next unit:

_____ Instructor's approval

_____ Progress plotted

UNIT 6

CAMS

OBJECTIVE

The student will construct cam displacement diagrams and cam profiles.

PRETEST

15-minute time limit

1. List four basic types of cam followers.

2. What is the function of a cam?

3. What determines the type of follower used?

4. What motion does a modified cam follower produce?

5. What are the two major kinds of cam designs?

6. What is a displacement diagram? What is it used for?

7. List three major kinds of cam motions.

8. Which of the three kinds of cam motions is the smoothest?

9. What is meant by dwell?

10. What is a working circle?

11. How are cams timed when more than one is on a given shaft?

12. In which direction is a cam laid out for clockwise rotation?

RELATED TERMS

Give a brief definition of each term as progress is made through the unit.

Kinds of followers _____

Dwell _____

Reciprocating motion _____

Working circle _____

Displacement diagram _____

Height _____

Time interval _____

Uniform velocity _____

Modified uniform velocity _____

Harmonic motion _____

Uniform acceleration _____

Rise/fall _____

Cam rotation direction _____

CAMS

A *cam* changes rotary motion. . . into up or down motion. . . . The cam is attached to a rotating shaft. The up and down, or *reciprocating*, motion is taken from the cam by the *follower*. Study the illustrations in figure 6-1. Note how the follower is at its highest point at 0 degree and 180 degrees and at its lowest point at 90 degrees and 270 degrees.

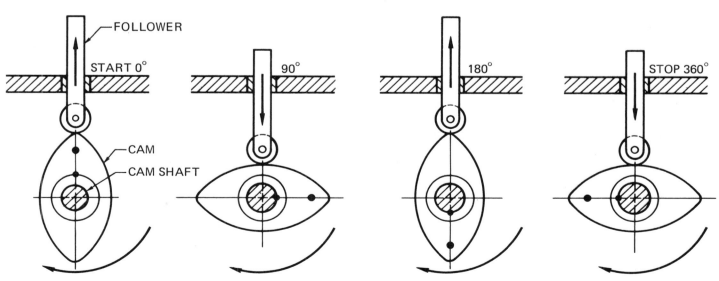

Fig. 6-1 Reciprocating motion from rotary movement

BASIC TYPES OF CAM FOLLOWERS

In selecting a *cam follower,* speed of rotation and the various loads placed upon the lifter must be considered, figure 6-2.

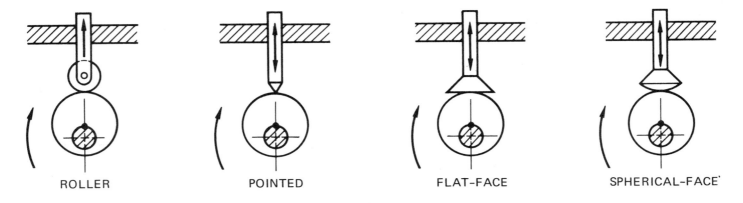

Fig. 6-2 Cam followers

Modified Cam Follower

The up and down motion can be modified by changing the rotary motion of shaft A, figure 6-3, into an up and down motion on follower B, and back into a rocker type motion on shaft C.

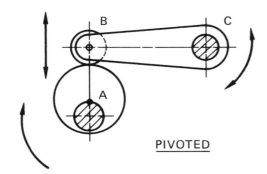

Fig. 6-3 Modified cam follower

CAM MECHANISM

There are two major kinds of cams, *radial design* and *cylindrical design.* The radial design can be modified to get a rocker motion, figure 6-4.

Fig. 6-4 Radial cam

Both designs use a *drive shaft* that rotates, but the action or direction of the followers differ. In radial arm design, the followers operate perpendicular to the cam shaft. In the cylindrical design, figure 6-5, the follower operates parallel to the cam shaft.

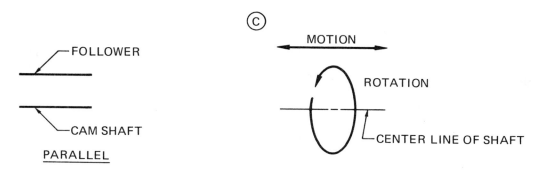

Fig. 6-5 Cylindrical cam

Figure 6-6 shows a cylindrical design. The shaft or rod that holds the follower does not rotate but moves back and forth. The follower produces movement parallel to the cam shaft.

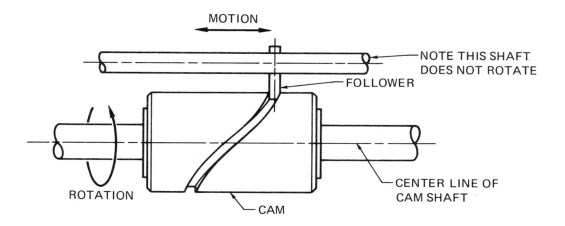

Fig. 6-6 Cylindrical cam. Follower produces movement parallel to cam shaft

DWELL

Dwell, or rest, is the time the cam follower is stopped or does not move up or down. The dwell is designed into the cam *profile.*

CAM DESIGN

In actual cam design, the cam follower type and location must be considered. For the drawing exercises in this unit, only cam design will be studied.

Study figure 6-7. Note that the cam rotates counterclockwise and the follower roller rotates *clockwise.* As the cam rotates, the follower will *drop* to location #2, then to #3, and so on. In designing a cam, it is important to consider the cam rotation direction. The cam is laid out and designed in the opposite direction of rotation.

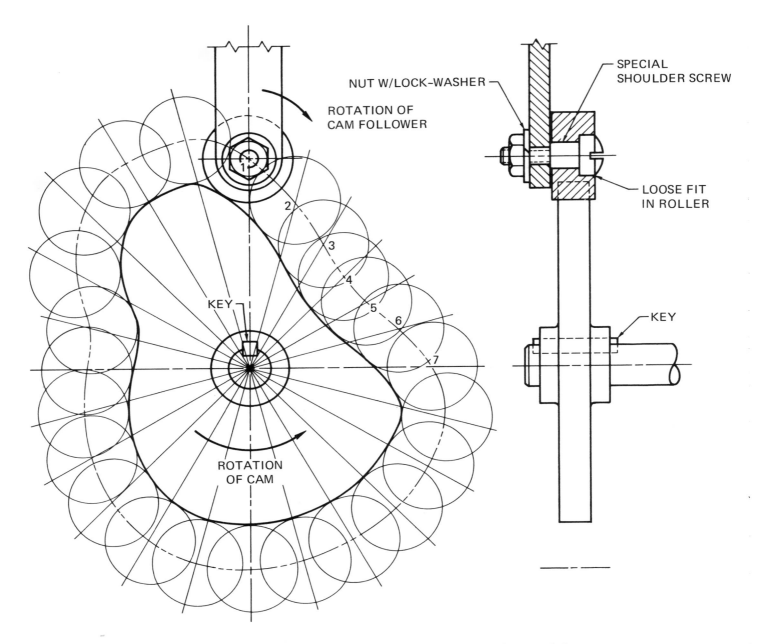

NUT W/LOCK-WASHER

ROTATION OF
CAM FOLLOWER

SPECIAL
SHOULDER SCREW

LOOSE FIT
IN ROLLER

KEY

KEY

ROTATION
OF CAM

Fig. 6-7 Cam rotation will cause the follower to drop and rise.

DISPLACEMENT DIAGRAMS

The *displacement diagram*, figure 6-8, is a curve that illustrates the exact motion of the follower through one full turn of the cam. The total overall length of the diagram does not have to be in scale or to exact dimension.

Terms of Displacement Diagrams

- *Length* equals one revolution of the cam. Length is drawn as a circumference of the work circle.
- *Working circle* is the radius equal to the distance from the center of the shaft to the highest point on the cam rise.
- *Height* is the maximum rise the follower travels. This is drawn to scale on the displacement diagram.
- *Time interval* is the time required for the cam to move the follower up or down.

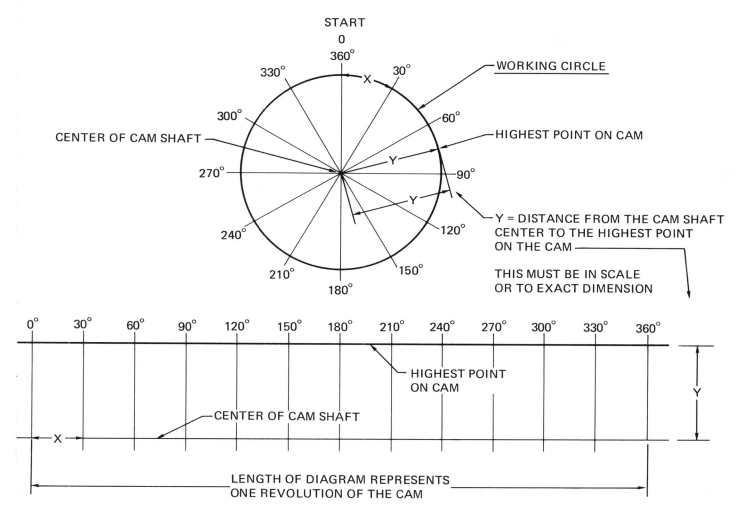

Fig. 6-8 Displacement diagram

CAM MOTIONS

There are three major kinds of cam motions; uniform velocity, harmonic motion, and uniform acceleration. Study how each is drawn.

Uniform Velocity

With uniform velocity, the follower rises and falls at a *constant speed.* This is very abrupt and rough, so it is usually modified by adding a 1/3 radius at the stop positions, figure 6-9.

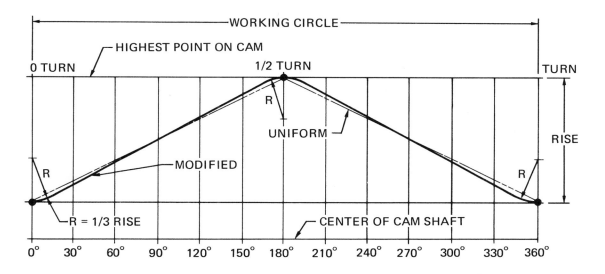

Fig. 6-9 Uniform velocity

Harmonic Motion

Harmonic motion is smooth, but the *speed is not uniform,* figure 6-10. It is used for high speed mechanisms as it has a smooth start and stop.

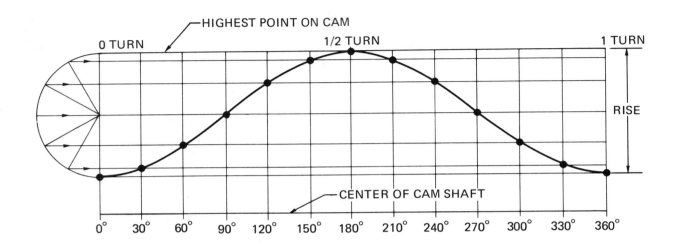

Fig. 6-10 Harmonic motion

Uniform Acceleration

Uniform acceleration is the smoothest of all cam motions. Its *speed is constant* throughout cam travel, figure 6-11.

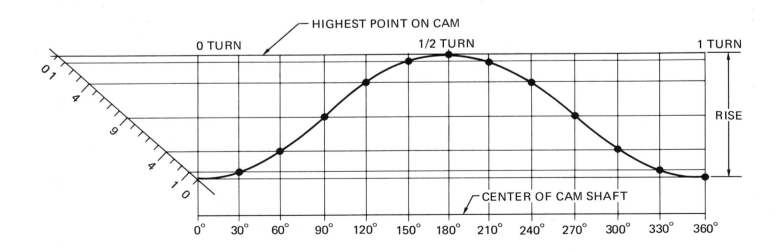

Fig. 6-11 Uniform acceleration

Practice Exercise 6-1

Using the displacement diagram for each cam motion, complete the cam profiles. Locate each point and draw straight lines from point to point. This is usually done with an irregular curve. Compare your work to the answer on page 000.

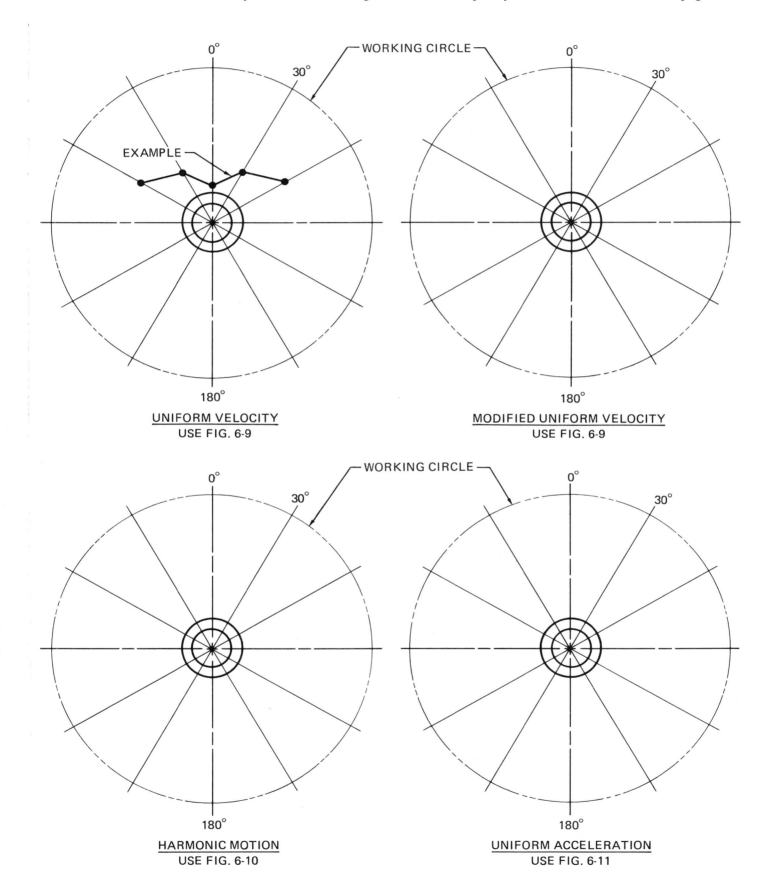

UNIFORM VELOCITY
USE FIG. 6-9

MODIFIED UNIFORM VELOCITY
USE FIG. 6-9

HARMONIC MOTION
USE FIG. 6-10

UNIFORM ACCELERATION
USE FIG. 6-11

CAM DESIGN AND LAYOUT

Study how to draw a cam layout from the given requirements to the displacement diagram.

Given:

1. Rise 90°, modified uniform velocity, .75″ (19)
2. Dwell 15°
3. Rise 90°, harmonic motion, .75″ (19)
4. Dwell 15°
5. Fall 150°, uniform acceleration, 1.5″ (38)
6. Counterclockwise

Carefully lay out the displacement diagram, figure 6-12.

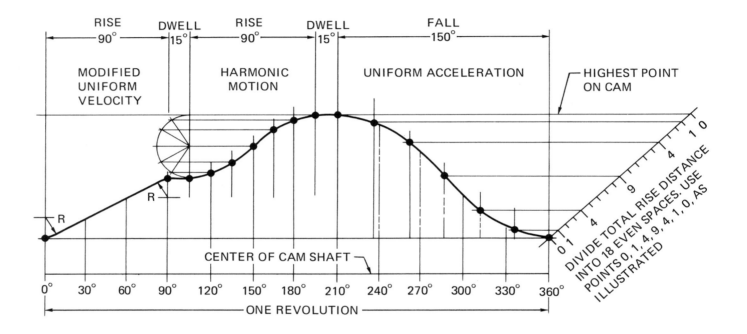

Fig. 6-12 Displacement diagram

Lay out cam. Always lay out cam points in directions opposite that of cam travel. In this example, figure 6-13, the cam rotates counterclockwise. Thus, lay out the cam clockwise as shown.

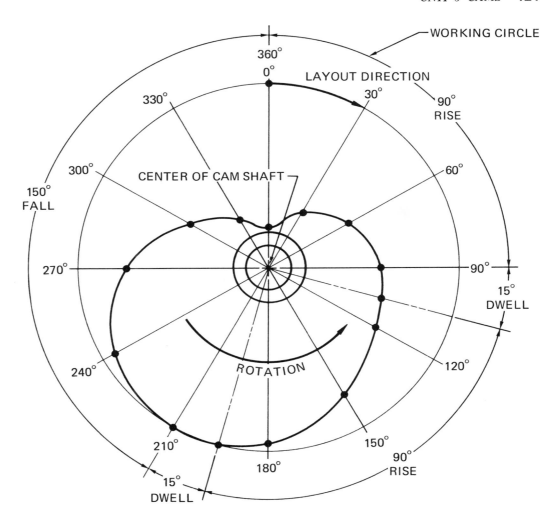

Fig. 6-13 Cam layout from given requirements

Practice Exercise 6-2

Draw a displacement diagram using the requirements given. Then, using the displacement diagram, lay out a cam of that diagram on the given working circle. The rotation is counterclockwise. Lay out the cam clockwise. Compare your work to the answer on page 129.

Given:

1. Rise 90°, modified uniform velocity, 1.5″ (38)
2. Dwell 60°
3. Fall 60°, modified uniform velocity, .75″ (19)
4. Dwell 30°
5. Fall 60°, modified uniform velocity, .75″ (19)
6. Dwell 60°

Displacement diagram:

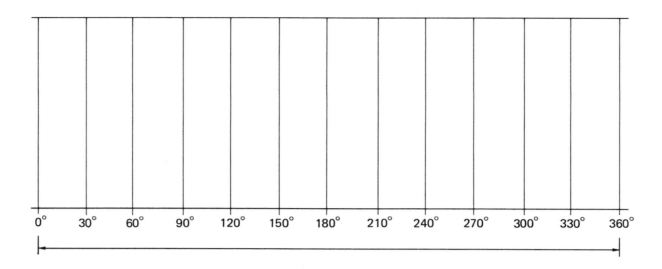

Working circle:

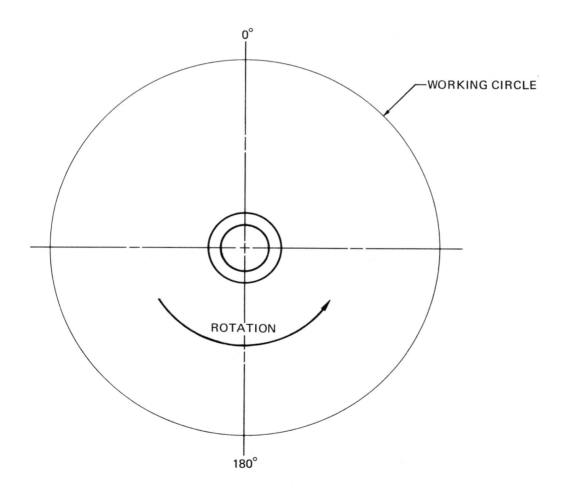

Practice Exercise 6-3

Draw a displacement diagram using the requirements given. Then, using the displacement diagram drawn, lay out a cam in the working circle. The cam rotation is clockwise. Compare your work to the answer on page 130.

Given:

1. Rise 90°, harmonic motion, 1.00″ (25)
2. Dwell 30°
3. Rise 90°, harmonic motion, .50″ (13)
4. Dwell 30°
5. Fall 90°, harmonic motion, 1.50″ (38)
6. Dwell 30°

Displacement diagram:

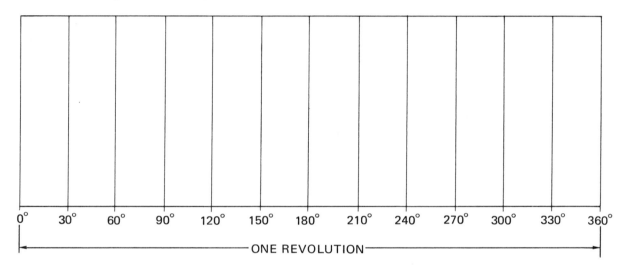

ONE REVOLUTION

Working circle:

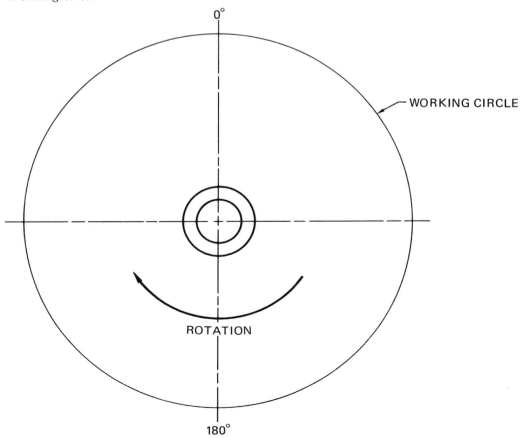

Practice Exercise 6-4

Draw a displacement diagram using the requirements given. Then, using the displacement diagram drawn, lay out a cam in the working circle. The cam rotation is clockwise. Compare your work to answer on page 131.

Given:

1. Fall 180°, uniform acceleration, 1.50″ (38)
2. Dwell 45°
3. Rise 90°, uniform acceleration, 1.50″ (38)
4. Dwell 45°

Displacement diagram:

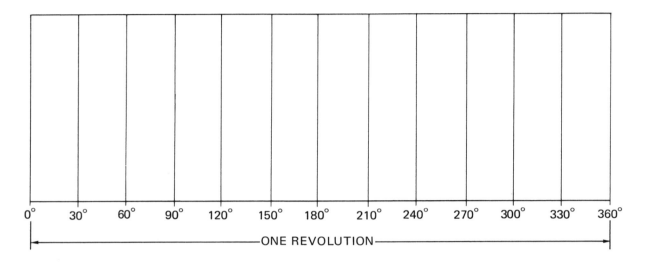

Working circle:

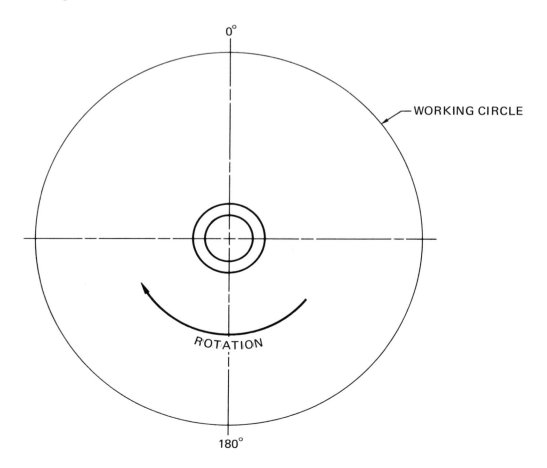

CAM TIMING

More than one cam is often attached to the same shaft. Each cam must function in relation to the other. Figure 6-14 shows the action of three cams attached to the same shaft.

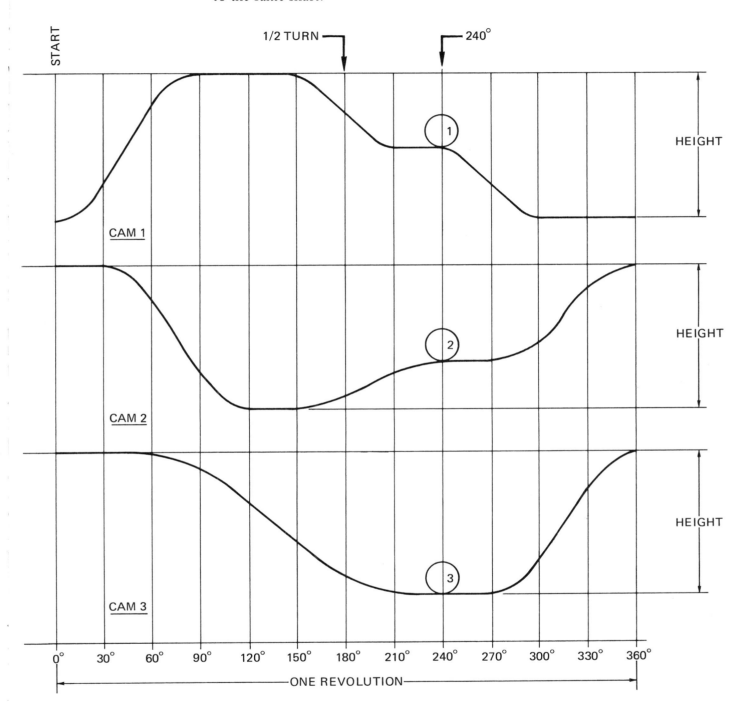

Fig. 6-14 Timing diagram

Notice cams 2 and 3 start at the highest point and cam 1 starts at the lowest point (0 degree). At 240 degrees, cam 1 is just completing dwell and is about to fall, cam 2 is just beginning a dwell, and cam 3 is halfway through a dwell. It is important these cams are designed with keyways so they cannot get out of time with each other.

Practice Exercises 6-5, 6-6, and 6-7

Draw a complete cam layout for each exercise using the given requirements. Draw a displacement diagram first. Do not dimension the drawings. Compare your work to the answers on pages 132 through 134.

Exercise 6-5

Given:

A working circle with a 7.5-inch (190) diameter hub, .875-inch (22) outside diameter, and .500-inch (12) inside diameter. Rotation direction is clockwise.

1. Rise 60°, harmonic motion, 1.00″ (25)
2. Dwell 15°
3. Rise 90°, harmonic motion, 1.50″ (38)
4. Dwell 15°
5. Fall 45°, harmonic motion, 1.75″ (45)
6. Dwell 15°
7. Rise 30°, harmonic motion, .75″ (19)
8. Dwell 15°
9. Fall to starting level 75°, harmonic motion, 1.5″ (38)

Exercise 6-6

Given:

A working circle with a 7-inch (178) diameter hub, .875-inch (22) outside diameter, and .500-inch (12) inside diameter. Rotation direction is couterclockwise.

1. Rise 120°, modified uniform velocity, 2.5″ (64)
2. Dwell 60°
3. Fall 30°, modified uniform velocity, .75″ (19)
4. Dwell 30°
5. Fall to starting level 90°, modified uniform velocity, 1.75″ (45)
6. Dwell 30°

Exercise 6-7

Given:

A working circle with an 8-inch (203) diameter hub, .875-inch (22) outside diameter, and .500-inch (12) inside diameter. Rotation direction is clockwise.

1. Fall 90°, uniform acceleration, 3″ (76)
2. Dwell 15°
3. Rise 60°, uniform acceleration, 2″ (50)
4. Dwell 105°
5. Rise 90°, uniform acceleration to starting level, 1″ (25)

DIMENSIONING CAMS

In order to dimension a cam, a displacement diagram must first be made to fit the requirements needed. The displacement diagram is then measured and dimensioned, figure 6-15.

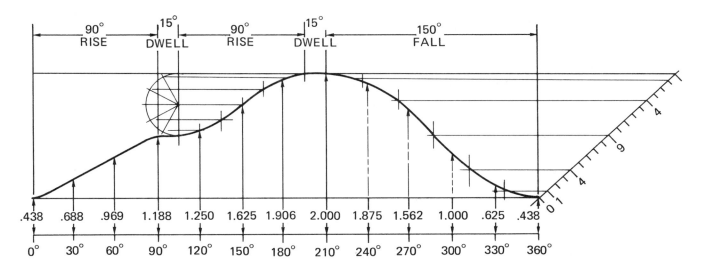

Fig. 6-15 Dimensioning displacement diagram

These dimensions are used to plot the points needed to lay out and dimension the cam. The cam in figure 6-16 is dimensioned from the center to the cam profile at every 30-degree segment of the working circle. Usually the rise, fall, and dwell areas are also included. This is only a suggested procedure as there is no standard for dimensioning cams. However, cams must be dimensioned neatly, completely, and accurately.

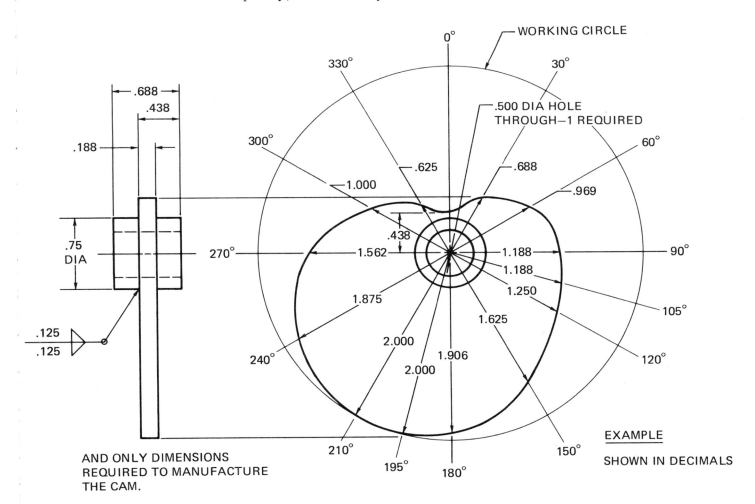

Fig. 6-16 Cam dimensioning

ANSWERS TO PRACTICE EXERCISES

Exercise 6-1

Carefully check each motion. Notice how smooth the uniform acceleration cam is in comparison to the uniform velocity cam. Remember that the cam is always finished with an irregular curve in order to have a smooth cam profile. (Note: drawings are not full size)

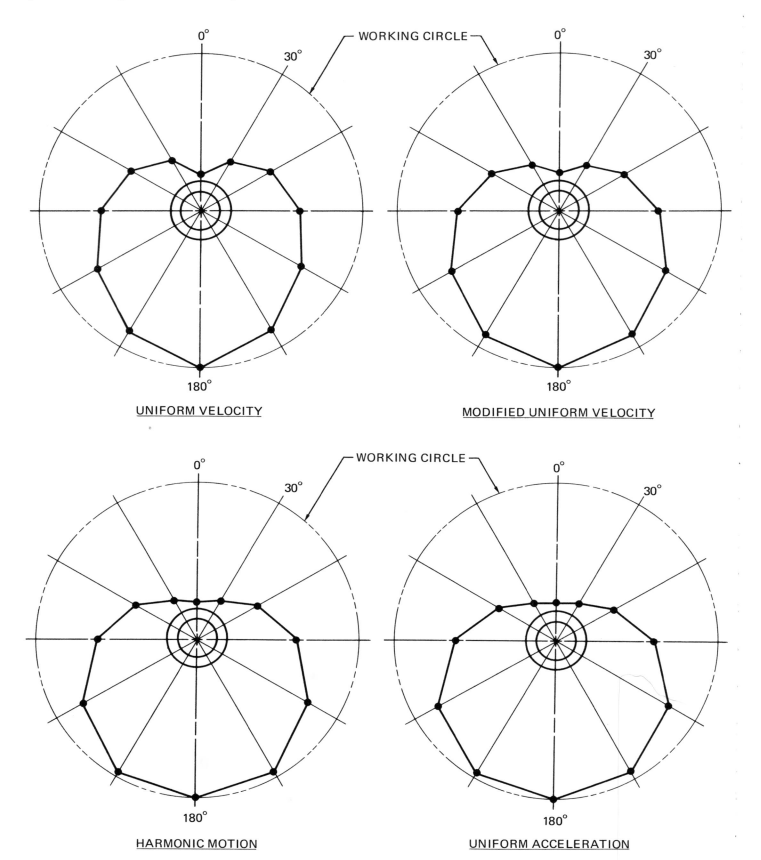

UNIFORM VELOCITY

MODIFIED UNIFORM VELOCITY

HARMONIC MOTION

UNIFORM ACCELERATION

Exercise 6-2

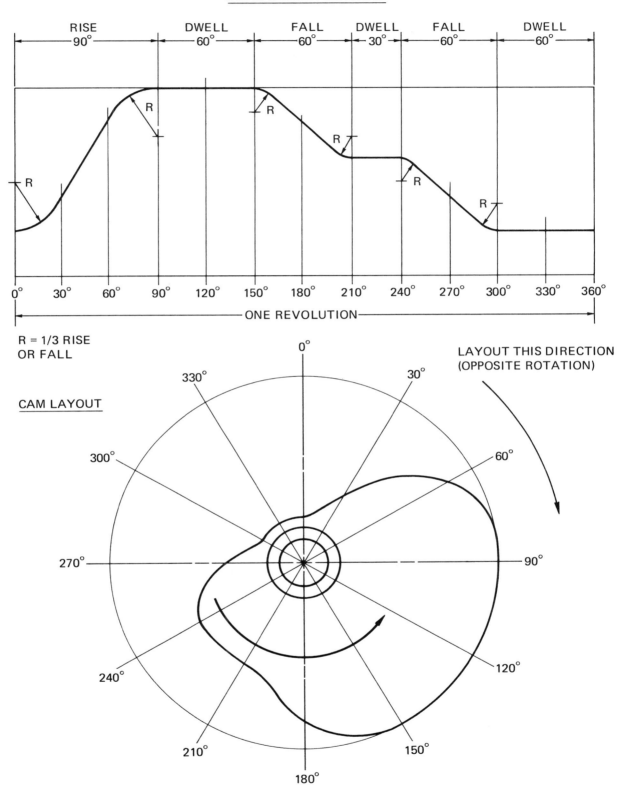

DISPLACEMENT DIAGRAM

R = 1/3 RISE
OR FALL

CAM LAYOUT

LAYOUT THIS DIRECTION
(OPPOSITE ROTATION)

Exercise 6-3

DISPLACEMENT DIAGRAM

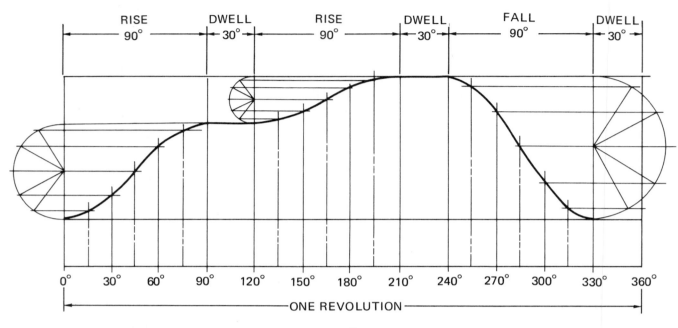

CAM LAYOUT

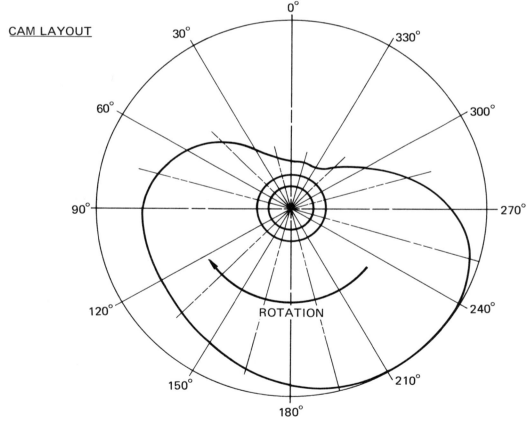

Exercise 6-4

DISPLACEMENT DIAGRAM

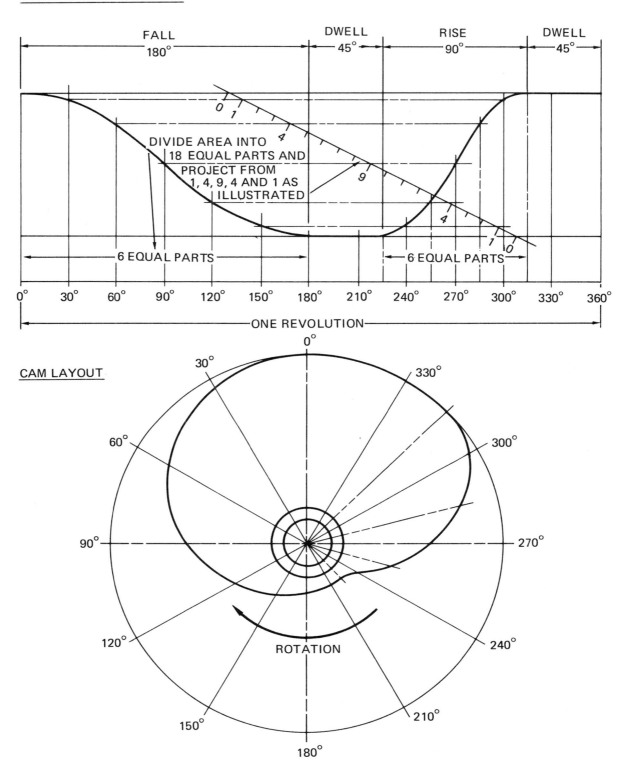

CAM LAYOUT

Exercise 6-5

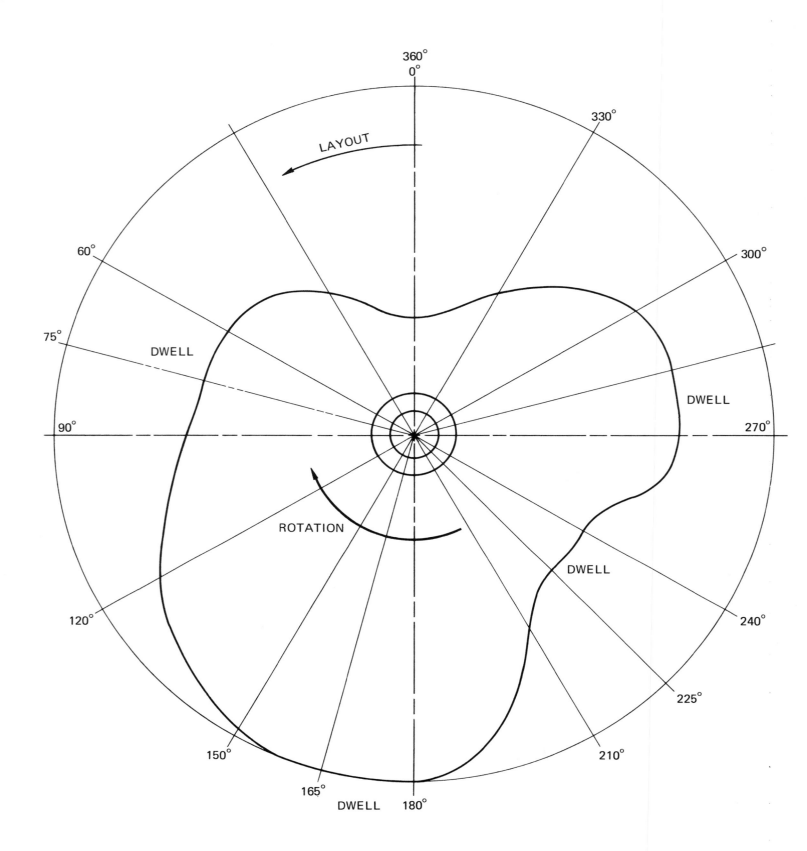

Exercise 6-6

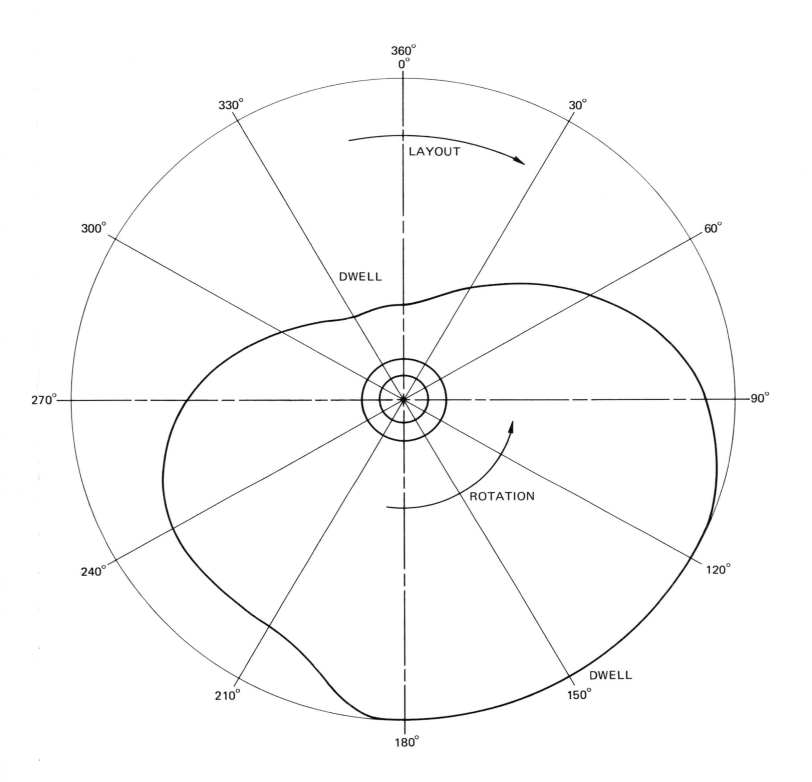

Exercise 6-7

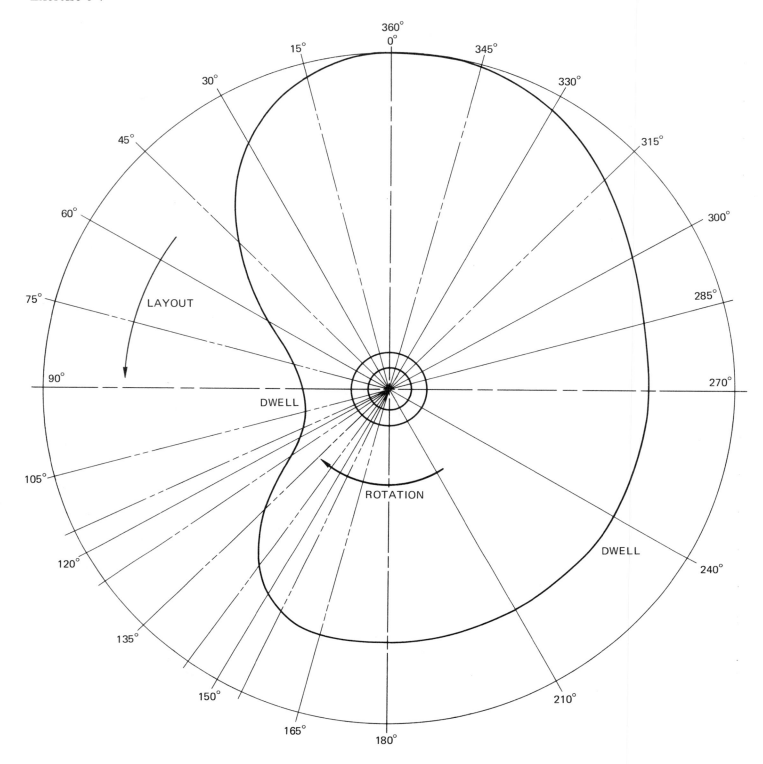

UNIT REVIEW

One-hour time limit

A. Lay out a displacement diagram using the given requirements:

1. Working circle 7.75″ (197) diameter
2. Rotation direction clockwise
3. Fall 75°, uniform acceleration, 2.25″ (57)
4. Dwell 15°
5. Rise 30°, harmonic motion, 1.25″ (32)
6. Dwell 5°

7. Fall 70°, modified uniform velocity, 2″ (50)
8. Dwell 15°
9. Rise 45°, harmonic motion, 1.75″ (44)
10. Dwell 30°
11. Rise 15°, uniform acceleration, 1.25″ (32)
12. Dwell 60°

Do not add dimensions. Total height is 3 inches (76)

B.   Lay out a cam using the displacement diagram drawn in A. Do not dimension.

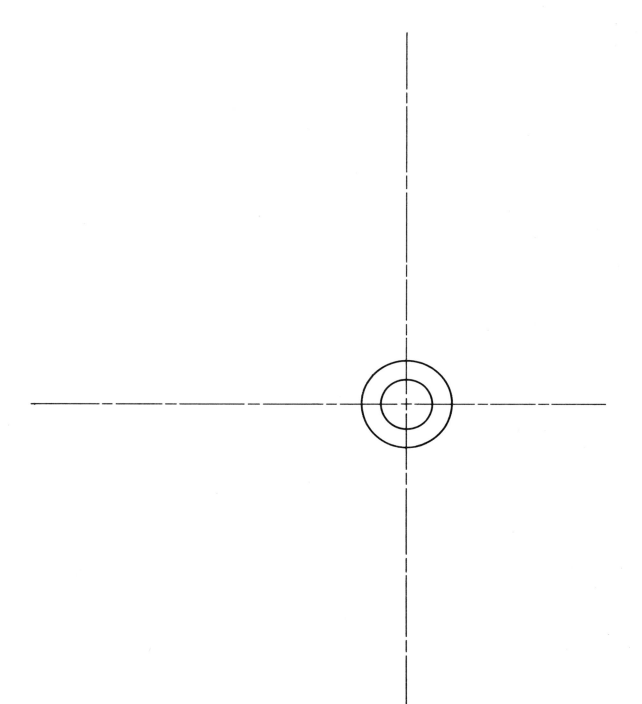

Before proceeding to the next unit:

_____ Instructor's approval

_____ Progress plotted

UNIT 7

ADVANCED DIMENSIONING

OBJECTIVE

The student will learn the various methods used to calculate limits, tolerances, and fits.

PRETEST

No time limit

1. How should a standard reamed hole be called out on a drawing?

2. Explain the difference between unilateral and bilateral tolerancing?

3. List three major kinds of fits.

4. What does T.I.R. mean? Where is it used?

5. What does the number 125 in a $\diagup$ mean?

6. What is tolerance?

7. What surface texture can be expected from a sand casting?

8. Using an RC-6 class of fit, what is the allowance between a 3/4-inch diameter (nominal size) shaft and hole?

9. What is the MMC of a hole? Of a shaft? Explain in full.

10. What shape tolerance zone results when using bilateral tolerancing? What shape tolerance zone results when using true positioning?

11. Using normal tolerances, is a hole with a 5/8-inch diameter the same size as a hole with a .625-inch diameter? Explain in full which hole costs more to produce.

12. Explain the term *limits*.

RELATED TERMS

Give a brief definition of each term as progress is made through the unit.

Limits _____

Tolerances _____

MMC _____

Allowance _____

Nominal size _____

Basic shape _____

Basic hole system _____

Clearance fit _____

Interference fit _____

Transitional fit _____

Unilateral tolerancing _____

Bilateral tolerancing _____

Tolerance buildup _____

Geometric tolerancing _____

Feature control symbol _____

Flatness _____

Straightness _____

Cylindricity _____

Roundness _____

Parallelism _____

Perpendicularity _____

Angularity _____

Concentricity _____

Symmetry _____

Modifiers (M and S) _____

TIR _____

True position (TP) _____

Tolerancing zone _____

DIMENSIONING

Dimensioning is perhaps the most important part of the drafter's job. It is even more important than speed, neatness, and accuracy. A drawing can be of poor quality but, if dimensioned correctly, could be used. The ideal drawing, one that the beginning drafter should strive for, is a drawing that is neat, in exact scale, dimensioned correctly, and completed in the shortest amount of time.

A fully qualified drafter must know and fully understand how the part will be manufactured, what tolerances should be applied, how the part will function with other parts, and how to describe or illustrate the part on paper so there is no misinterpretation by the worker who will make the part.

No two parts can be made exactly alike, but they can be made within specific tolerances so they are interchangeable. Because of this fact, a tolerance system was devised which is a part of the national standard for dimensioning. It is the drafter's job to understand and use this system of tolerancing. All illustrations in this unit are in fractions and/or decimals. The metric system uses exactly the same process.

LIMITS AND TOLERANCE

Think of the signs along the interstate highways of our country. They tell motorists how fast or slow they are allowed to drive. If they go faster than posted, they could be fined for speeding. If they drive slower than posted, they could also be fined, figure 7-1.

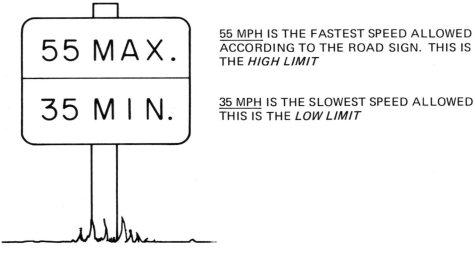

55 MPH IS THE FASTEST SPEED ALLOWED ACCORDING TO THE ROAD SIGN. THIS IS THE *HIGH LIMIT*

35 MPH IS THE SLOWEST SPEED ALLOWED THIS IS THE *LOW LIMIT*

Fig. 7-1 Driving tolerances

The *tolerance* in the example in figure 7-1 is the difference between the high limit and the low limit. Thus:

$$
\begin{array}{ll}
55 \text{ mph} & \text{high limit} \\
-35 \text{ mph} & \text{low limit} \\
\hline
20 \text{ mph} & \text{accepted } tolerance
\end{array}
$$

The drafter must state the largest and the smallest size hole that is acceptable for a particular application or function.

In figure 7-2, .505 inch is the *largest* hole allowed, the high limit; .500 inch is the *smallest* hole allowed, the low limit. The tolerance in this example is the difference between the high limit and the low limit. Thus:

$$
\begin{array}{ll}
0.505'' & \text{diameter (high limit)} \\
-0.500'' & \text{diameter (low limit)} \\
\hline
0.005'' & \text{design tolerance}
\end{array}
$$

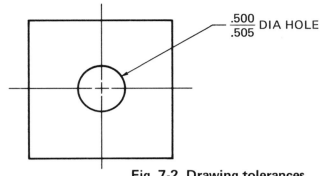

.500 / .505 DIA HOLE

Fig. 7-2 Drawing tolerances

.500 / .505 DIA HOLE

Fig. 7-3 Tolerance of a hole

Never call for closer or tighter limits than are necessary. Closer or tighter limits than necessary are more costly to produce. Try to design for the best function at minimum cost. Some things to consider when determining tolerances include:

- Length of time parts are engaged
- Speed (if any) mating parts will move or turn
- Lubrication
- Temperature & humidity
- Material used
- Estimated "life" required
- Capability of company to produce the tolerance
- Cost (very important)

In calling out a hole, place the smallest limit on top, figure 7-3. The theory for doing this is, if a machinist tries for the top figure (.500") and the hole is too small, it can be redrilled larger. If it is drilled a little larger than the top limit (smallest), it will still fall within the given tolerance.

In calling out a shaft, place the largest limit on top, figure 7-4. The theory for doing this is, if a machinist tries for the top figure (.495") and makes the shaft too large, it can be machined smaller. If it is machined a little smaller than the top limit (largest), it will still fall within the given tolerance. Note that various company standards differ, but for *all* problems done in this course use the standard presented here.

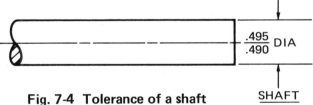

.495 / .490 DIA

Fig. 7-4 Tolerance of a shaft   SHAFT

Maximum Material Condition

If an object with a hole size of .500-inch diameter was placed on a scale, it would weigh *more* than the same size object with a .505-inch diameter hole drilled in it, figure 7-5. The *maximum material condition* (MMC) is the smallest size the hole can be.

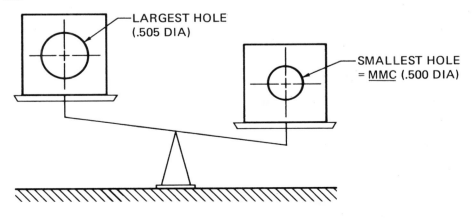

LARGEST HOLE (.505 DIA)

SMALLEST HOLE = MMC (.500 DIA)

Fig. 7-5 MMC of a hole

If a shaft with a .495-inch diameter is placed on a scale, it would weigh *more* than a shaft with a .490-inch diameter. The *maximum material condition* (MMC) is the largest size the shaft can be, figure 7-6.

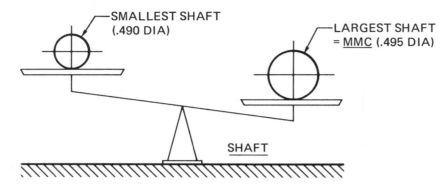

Fig. 7-6 Maximum material condition of a shaft

Thus, the *maximum material condition* is the condition where a feature of size contains the maximum amount of material within the stated limits of size.

ALLOWANCE OR CLEARANCE

Allowance is the intentional difference between the sizes of mating parts. It is expressed as either clearance (+) allowance or interference (−) allowance. Figure 7-7 shows how to find the minimum clearance (+) allowance, and figure 7-8 shows how to find the maximum clearance (+) allowance.

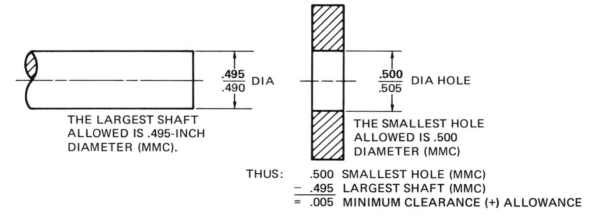

Fig. 7-7 Derivation of minimum clearance allowance

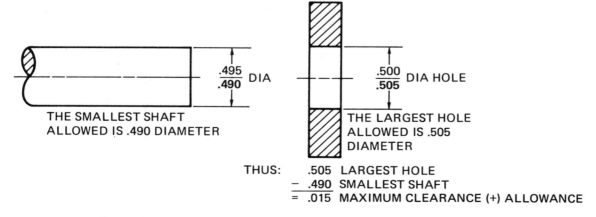

Fig. 7-8 Derivation of maximum clearance allowance

Nominal Size

Nominal size is the designation used for general identification only. For example, a steel plate that is referred to as 1/4 inch (6) thick. If measured, however, it is actually more or less than 1/4 inch (6) thick.

Basic Size

Basic size is the size the drafter starts with before applying the required limits to it. There are two systems: the *basic hole system* and the *basic shaft system*. The basic hole system starts with the hole size (basic size) and adjusts the shaft size to fit. The latter system starts with the shaft size (basic size) and adjusts the hole size to fit. Because holes usually are made with standard tools (drills, reams, bores) it is best to use the basic hole system. The shafts can be made to most any size with little problem.

How to Determine Size Using the Basic Hole System

Use the basic hole system for all problems in this unit. Starting with the nominal size, determine the basic size (smallest hole), and calculate all sizes needed. Label each item as illustrated. Dimensions must be correct. Remember to dimension holes with the smallest limit on top, and the shaft with the largest limit on top, figure 7-9.

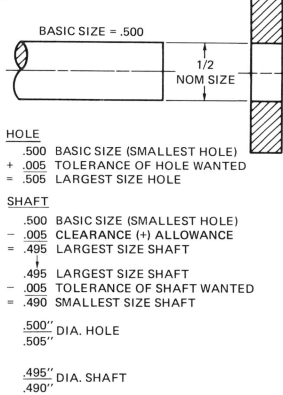

Carefully transfer calculated values to the drawing:

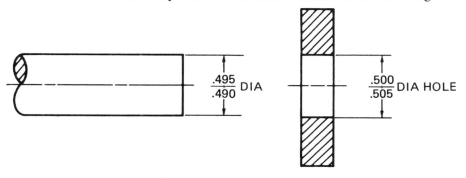

Fig. 7-9 Using basic hole system

KINDS OF FITS

A *fit* is a general term used to signify the range of tightness or looseness which results from the application of a specific combination of allowances and tolerances in mating parts. There are many kinds of fits. However, only a limited number will be discussed in this unit.

A *clearance fit* (+) has limits of size that result in a clearance when mating parts are assembled. The parts can be assembled by hand because the hole is always larger than the shaft.

An *interference fit* (–) has limits of size that result in an interference when mating parts are assembled. Parts must be pressed together because the hole is always smaller than the shaft.

In a *transition fit* (+ or –), the limits of size result in either a clearance or interference when mating parts are assembled.

Practice Exercise 7-1

Place each answer in the space provided. Indicate if the fit is plus (+) or minus (–). Remember: Dimensions of holes — smallest limit on top; dimensions of shafts — largest limit on top. A positive allowance = clearance. A negative allowance = interference. Compare your work to the answers on page 178.

Remember:

```
          SMALLEST HOLE — Lower limit hole — (Basic Size)
     +      TOLERANCE
     =    LARGEST HOLE — Upper limit hole
```

(Allowance / Clearance fit) ← — or — → (Allowance / Interference fit)

SMALLEST HOLE	SMALLEST HOLE
– CLEARANCE ALLOWANCE	+ INTERFERENCE ALLOWANCE
= LARGEST SHAFT — *Upper limit* shaft	= LARGEST SHAFT — *Upper limit* shaft
LARGEST SHAFT	LARGEST SHAFT
– TOLERANCE	– TOLERANCE
= SMALLEST SHAFT — *Lower limit* shaft	= SMALLEST SHAFT — *Lower limit* shaft

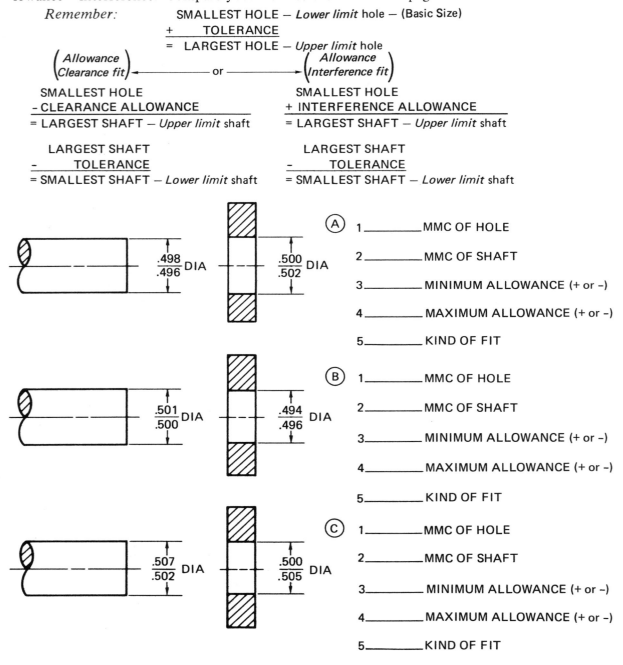

(A)
1. _____ MMC OF HOLE
2. _____ MMC OF SHAFT
3. _____ MINIMUM ALLOWANCE (+ or –)
4. _____ MAXIMUM ALLOWANCE (+ or –)
5. _____ KIND OF FIT

(B)
1. _____ MMC OF HOLE
2. _____ MMC OF SHAFT
3. _____ MINIMUM ALLOWANCE (+ or –)
4. _____ MAXIMUM ALLOWANCE (+ or –)
5. _____ KIND OF FIT

(C)
1. _____ MMC OF HOLE
2. _____ MMC OF SHAFT
3. _____ MINIMUM ALLOWANCE (+ or –)
4. _____ MAXIMUM ALLOWANCE (+ or –)
5. _____ KIND OF FIT

Practice Exercise 7-2

Fill in all of the blank spaces. Show the math in the space provided. Label each answer. Compare your work to the answers on pages 178 and 179.

Prob.	Nom. Size	Basic Size	Dim. of Hole	Dim. of Shaft	Hole Tol.	Shaft Tol.	Min. Allowance	Max. Allowance
1			.625 .630			.005	+ .002	
2	1 1/4				.001	.002	+ .001	
3			.312 .314	.316 .314				
4	3/4				.003	.002	+ .001	
5			.812 .813	.815 .814				

Do math below:

Keys and Slots

The limits, tolerances, and allowances for other close-fitting parts are determined the same way as for holes and shafts. In practice exercise 7-2, the slot takes the place of the hole, and the key takes the place of the shaft. All terms associated with hole and shaft apply.

Practice Exercise 7-3

Place each answer in the space provided. Indicate if it is a plus (+) or minus (–) fit. Remember, as in hole and shaft limits, the smallest slot limit goes on top, and the largest key limit goes on top. Positive allowance = clearance, and negative allowance = interference. Compare your work to the answers on page 179.

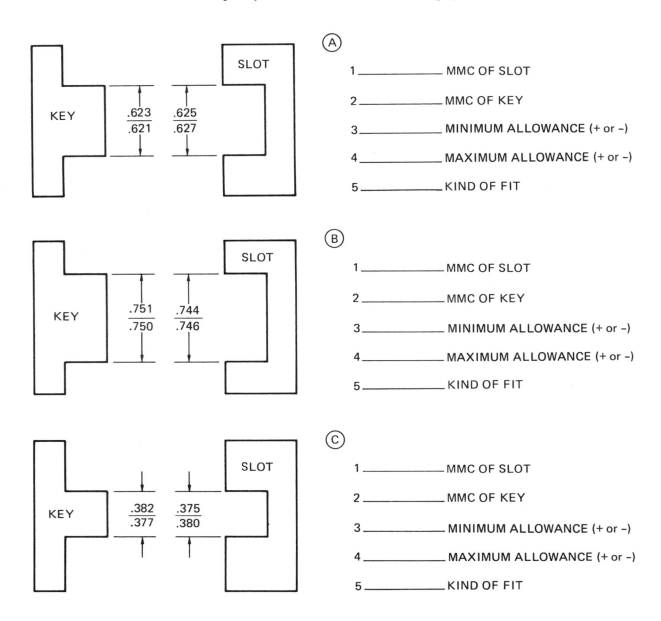

Ⓐ

1 _____ MMC OF SLOT

2 _____ MMC OF KEY

3 _____ MINIMUM ALLOWANCE (+ or –)

4 _____ MAXIMUM ALLOWANCE (+ or –)

5 _____ KIND OF FIT

Ⓑ

1 _____ MMC OF SLOT

2 _____ MMC OF KEY

3 _____ MINIMUM ALLOWANCE (+ or –)

4 _____ MAXIMUM ALLOWANCE (+ or –)

5 _____ KIND OF FIT

Ⓒ

1 _____ MMC OF SLOT

2 _____ MMC OF KEY

3 _____ MINIMUM ALLOWANCE (+ or –)

4 _____ MAXIMUM ALLOWANCE (+ or –)

5 _____ KIND OF FIT

TOLERANCING

Unilateral Tolerancing

Unilateral tolerancing means applying the tolerance in only one direction from the size actually wanted. Because parts cannot be made exactly alike, the drafter sometimes allows a larger tolerance than the exact size (dimension), but not smaller. This is called unilateral tolerance, figure 7-10.

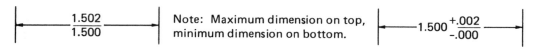

Fig. 7-10 Unilateral tolerance

Bilateral Tolerancing

Bilateral tolerancing means applying a portion of the tolerance in two directions from the size actually wanted. For example, a drafter knows that an exact 1.500-inch length cannot be made, so + .002 or –.002 inch from the size required is tolerated. This is called a bilateral tolerance, figure 7-11.

Fig. 7-11 Bilateral tolerance

Tolerance Allowance Chart

In drawing forms used by many companies, the title block includes a tolerance chart, such as the one in figure 7-12.

Tolerance unless otherwise spec.	
Fractions +/-.015 *(1/64)*	
.XX	+/-.015
.XXX	+/-.010
.XXXX	+/-.0002

Fig. 7-12 Tolerance allowance chart

(Note: *"unless otherwise spec."* means any dimension that must be held closer will have limits directly on it.)

Using the tolerance chart in figure 7-12, any fractional dimension will automatically have a ± .015 inch (1/64) tolerance applied to it, any two-place decimal (.xx) will automatically have a ± .015-inch tolerance applied to it, etc. Using a 1/2-inch nominal size as reference:

CALLOUTS FOR HOLES

Each manufacturing company has its own standard for calling out hole dimensions. Figure 7-13 illustrates the callout standard most frequently used in industry. The drafter does not specify on a drawing the process to be used but merely indicates the limits required of the hole. Specify the number of holes required, as illustrated, even if only one hole is made.

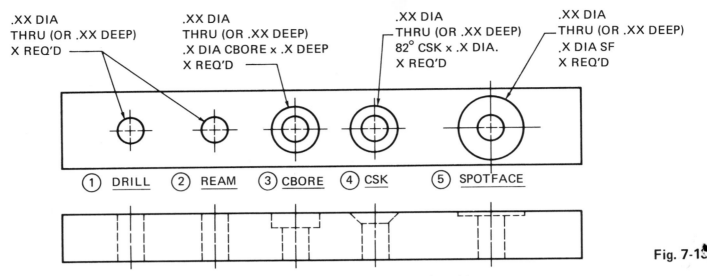

Fig. 7-13

1. Drilling is used most often to make holes. The *drilled hole* produced is acceptable unless a limit is indicated. If it is not clear on the drawing, the notation also indicates through holes or the depth of the hole, and the number required.
2. A *reamed hole* is called out the same as a drilled hole, except it will always have limits which can be made only with a reamer.
3. A *counterbored hole* must include hole diameter, diameter and depth of the counterbore and the number required.
4. A *countersunk hole* must include the hole diameter, angle of countersink (82-degree angle is standard), the diameter of countersink measured across the top of the countersunk hole, and the number required.
5. A *spotface* must include hole diameter, hole depth, spotface diameter, and number required. Do not call out spotface depth as it varies depending upon the surface irregularities. A spotface is made only to that depth necessary to produce a flat surface or a predetermined thickness recorded for a feature; i.e. a flange.

DRAWING LIMITS AND TOLERANCES

Figure 7-14 shows a drawing of the shaft size desired by the drafter. Figure 7-15 shows the same drawing with limits and tolerances added to the desired shaft size.

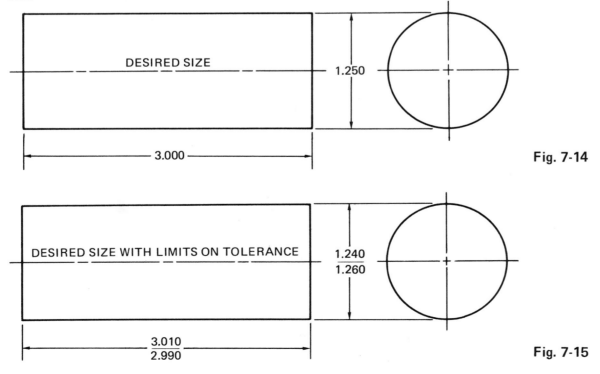

Fig. 7-14

Fig. 7-15

The drawing in figure 7-16 shows the size range in which the shaft must fall to be accepted by the inspection department. If the shaft was machined to a length less than 2.990 inch, it would be scrapped. If it was over 3.010 inches in length, it would be remachined to fit within given limits and tolerances.

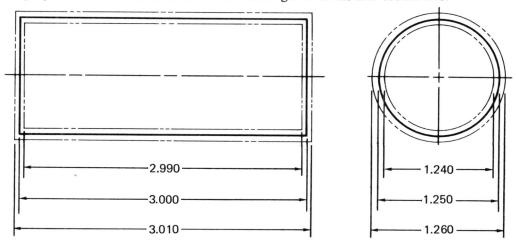

Fig. 7-16

The drawing in figure 7-17 shows the hole location desired by the drafter. Figure 7-18 is the same drawing with the hole location limits and tolerances added to desired hole location dimensions.

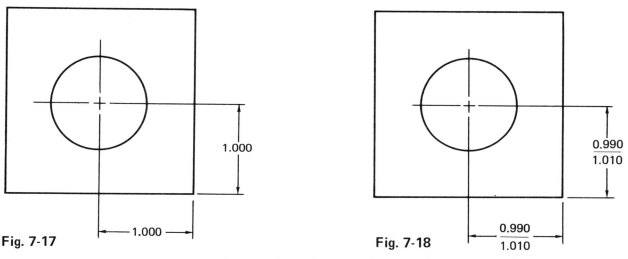

Fig. 7-17

Fig. 7-18

Figure 7-19 is the same drawing showing the effect of the hole location using acceptable limits and tolerances. If the hole location varies from the acceptable location limits and tolerances, the inspection department will scrap the part.

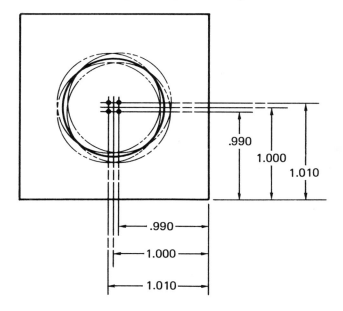

Fig. 7-19

TOLERANCE ERROR

A *buildup of tolerance* will result in tolerances much larger than those that are acceptable. Perform the following experiment to learn what tolerance buildup refers to.

1. Set a compass with a sharp lead to an exact opening of 11/16 inch.
2. Step off 10 equal 11/16-inch spaces starting at the point noted by "start" in figure 7-20.
3. Measure the total overall length stepped off.

START

Fig. 7-20

The length should be 6 7/8 inches (10 × 11/16"). If the answer is off it is because the original compass setting was inaccurate (a little long or short of 11/16 inch), and that inaccuracy was multiplied ten times. This type of error is called a buildup of tolerances.

The tolerance of one-place decimals in figure 7-21 is given as ± .015 inch. If the machinist makes the object using all long dimensions, as shown in the lower view, the object length totals 4.075 inches. This is considerably longer than the acceptable length of 4.015 inches. This occurred because of a tolerance buildup. The same situation would occur if the lower limits were used.

AS DRAWN AND DIMENSIONED

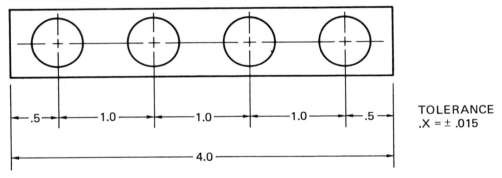

TOLERANCE
.X = ± .015

AS MANUFACTURED

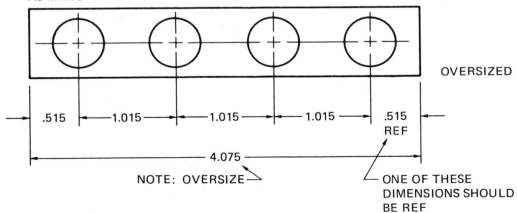

OVERSIZED

NOTE: OVERSIZE

ONE OF THESE DIMENSIONS SHOULD BE REF

Fig. 7-21

Figure 7-22 shows how a tolerance buildup is avoided. If numerical control machines are used, there will not be any problem with incremental dimensioning in figure 7-21. This is because the inherent accuracy of the machines is closer than the specified limits.

Base Line Dimensioning

Base line dimensioning is the best solution to tolerance buildup. Using the left side (or any important edge, point, or center line as a feature) as a base or reference, project all dimensions from the same reference base.

AS DRAWN AND DIMENSIONED

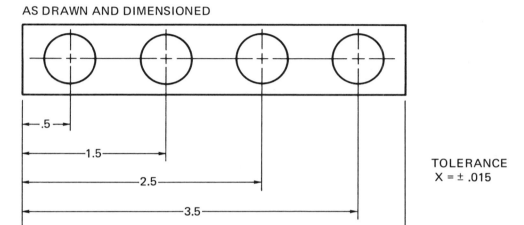

AS MANUFACTURED

SHOWN WITH ALL DIMENSIONS
MANUFACTURED OVERSIZE
WITHIN SPECIFICATION

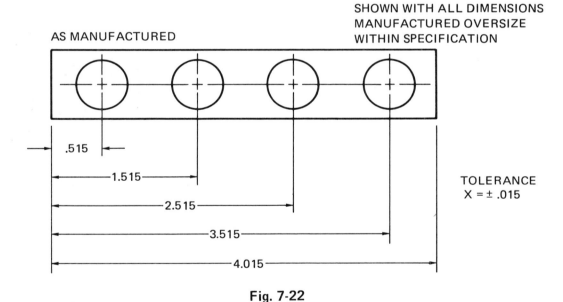

Fig. 7-22

The drafter must consider the possibility one part could be made on the minus side of given tolerance and another could be made on the plus side of the factured, with all dimensions oversize but within specifications. The worse condition is either .015 inch oversize or .015 inch undersize. There is no chance of a tolerance buildup using this method of dimensioning.

The drafter must consider the possibility part 'A' could be made on the minus side of given tolerance and part 'B' could be made on the plus side of the given tolerance. The drafter should consider carefully where to place the dimensions, which dimensions should have close tolerances and which should be very loose, and, above all, how various mating parts go together. It is important that mating parts be dimensioned from the same base line or feature so that a buildup of tolerances cannot occur.

Study figure 7-23. Note that the holes in part 'A' must be large enough to allow for hole location tolerance, location tolerance, and size tolerance in part 'B'. Notice that even though an edge was used as a base line for dimensioning other holes, the base line could also have a *hole* from which other holes would be dimensioned. The important consideration is determining from the design which characteristic is the most important.

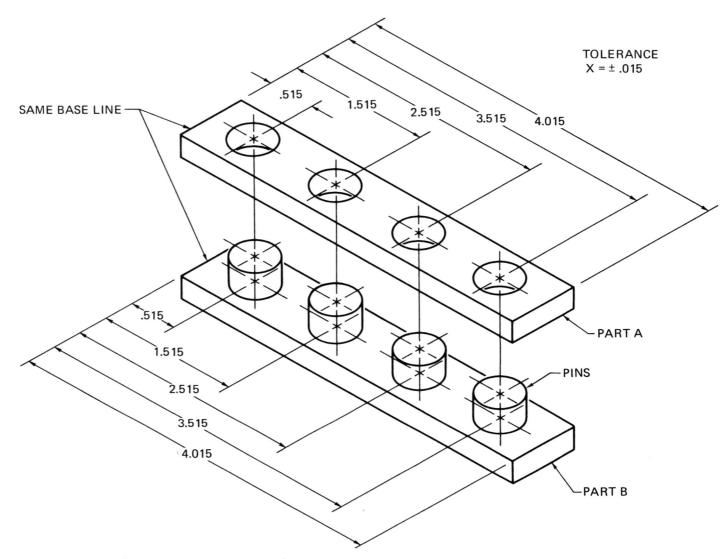

Fig. 7-23 Base line dimensioning

Other factors to consider in deciding the tolerances are:

- Length of time parts are engaged
- Speed at which mating parts move
- Load on part (s)
- Lubrication used (if necessary)
- Average temperature at which parts operate
- Humidity factor
- Materials from which parts are made
- Estimated "life" required of part
- Capability of machinery, tools, and machinist manufacturing part
- Cost

Practice Exercise 7-4

Study the detail drawing. Note how important line work is. The object should stand out and the dimensions should be properly placed. Thought must be given to dimensioning a drawing correctly. Notice how blind holes are illustrated.

Answer questions 1 through 12 in the space provided on pages 154 and 155 using the detail drawing. Compare your work to the answers on pages 179 and 180.

1. Having carefully studied the detail drawing, list the controlling feature which is the most important. Why?

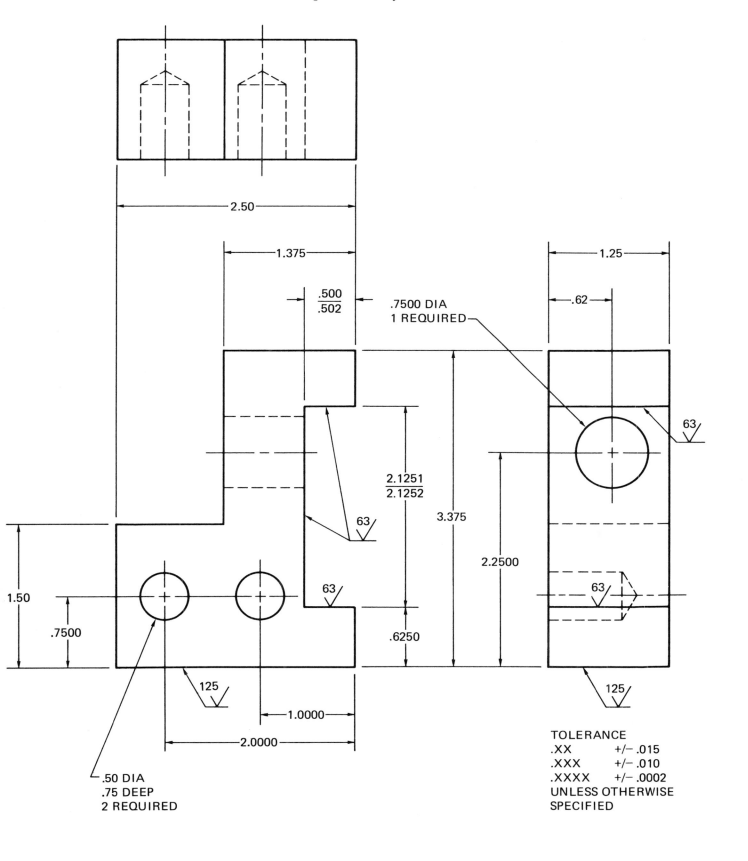

2. What surface finish is called for on the groove?

3. Which process is used to manufacture the .50-inch diameter holes? Why?

4. Which process is used to manufacture the .7500-inch diameter hole? Why?

5. Why is the bottom surface used as a reference line for vertical decimals?

6. What is the maximum and minimum limits for locating the height of the .50-inch diameter holes above the reference line?

7. How close to each other can the two .50-inch diameter holes be drilled? Explain.

8. Which kind of a dimension is the .500/.502-inch dimension on the groove depth?

9. What are the maximum and minimum limits the .7500-inch diameter hole can be from the reference line? What is the maximum tolerance?

10. What is the tightest limit in the drawing?

11. Where are the two-place decimals used? (2.50″, 1.50″, 1.25″, .62″, and .50″ diameter)

12. The distance from the base line to the maximum height of the groove could be _____ and still be within the tolerance.

TYPES AND USES OF FITS

A standard system of fits is used to help manufacturers, for example, use fewer gauges and to provide the optimum design control. As a consequence, interchangeable parts are more readily achieved. Standard fits are noted in figure 7-24 with an explanation of where each is used. The drafter must choose as loose a fit as applicable to the design requirements. Unnecessarily close fits that are not required by the design are more costly to manufacture.

Sym.	Type of Fit	No.	Usage
RC	Running & Sliding	1	Accurate location — little or no "play."
		2	Accurate location—more clearance-turn or move at slow speed.
		3	Precision running at slow speed-light pressure-no temp. change.
		4	Accurate running at moderate speed-medium pressure-min. "play."
		5	High speed — heavy pressure.
		6	
		7	Free running — not accurate — large temperature changes.
		8	Loose — not accurate — low price.
		9	
LC	Locational Clearance	1	Normally stationary parts — easily assembled/disassembled.
		2	
		3	
		4	
		5	
		6	
		7	
		8	
		9	
		10	
		11	
LT	Transitional Locational	1	Close accuracy of mating parts — little or no interference.
		2	
		3	
		4	
		5	
		6	
		7	
LN	Locational Interference	2	Exact accuracy of mating parts.
		3	
FN	Force & Shrink	1	Light drive fit-permanent assembly — usually used with cast iron.
		2	Medium drive fit — or shrink fit — used for steel-small parts.
		3	Heavy drive fit — medium size parts — used for steel.
		4	Force fit or shrink fit in parts which can be highly stressed.
		5	

Fig. 7-24 Fits, kinds and uses

Choosing a Class of Fit

A drafter who is designing an assembly of 1 1/2-inch nominal size diameter shaft that must turn at a high speed under a heavy load or pressure must calculate tolerances on the hole and the shaft. In looking at figure 7-24, an RC–5 or RC–6 fit comes closest to the required specifications.

How to Figure Tolerance

(The inch system is used as an example, metric system is calculated exactly the same way)

Use A.S.M.E. # B4.1 chart, Appendix C.

1. The basic hole system is used in this method.
2. All limits on the chart are in 1000ths of an inch.

   For example:

   - 2.0 would mean 2 thousandth or .002 (move decimal point three places to the left).
   - 1.5 would mean 1 1/2 thousandths or .0015 (move decimal point three places to the left).
   - .8 would mean .8 thousandths or .0008 (move decimal point three places to the left).

3. Be sure to observe the plus (+) or minus (–) sign before each figure. All clearance fits must use minus (–) and all interference fits must use plus (+) to calculate the largest shaft. This is important!
4. Accuracy is a must. Check and double check all work.

   EXAMPLE:

   Carefully do the following steps:

   1. 1 1/2 inches (basic size) required (1.500").
   2. Locate either RC–5 or RC–6 (RC–6 will be used as an example) on the Running/Sliding Chart, page 311.
   3. Locate at the left of the chart, the range of size that 1 1/2 inches falls into (1.19 to 1.97).
   4. Calculate the limits of the hole using figures from the chart for "Hole Tolerance"

      1.5000 smallest hole-lower limit (Basic size)
      + .0025 hole tolerance
      1.5025 largest hole-upper limit

   5. Calculate the limits of the shaft:

      1.5000 smallest hole (basic size)
      – .0020 clearance allowance (from chart)*
      1.4980 largest shaft (upper limit, shaft)

      *Note — clearance allowances are *subtracted* (–)
          interference allowances are *added* (+)

      1.4980 largest shaft (upper limit, shaft)
      – .0016 tolerance of shaft
      1.4974 smallest shaft (lower limit, shaft)

   6. Check all math

This math procedure is the same as illustrated on page 144 using the basic "Hole System."

It should be noted that standard fits will apply in the majority of cases. However, when design requirements cannot be met with standard fits, use the same method for calculating the desired fit.

Practice Exercise 7-5

Very carefully determine the tolerances for the following problems. Use Appendix C. Double check all work, then compare your work to the answers on pages 180 and 181.

1. RC-8
    1 1/2″ = Dia.
    (38.1 MM)

Metric		Inch	
38.100		1.5000	smallest hole (basic size)
+ .100		+ .0040	hole tolerance
38.200		1.5040	largest hole
38.100		1.5000	basic size
− .130		− .0050	allowance (+ or −)
37.970		1.4950	largest shaft
37.970		1.4950	largest shaft
− .060		− .0025	shaft tolerance
37.910		1.4925	smallest shaft

2. LC-2
    5/8″ = Dia.
    (15.8750 MM)

3. LN-3
    3.0″ = Dia.
    (76.2 MM)

4. RC-3
    1 7/16″ = Dia.
    (36.5125MM)

5. LT-4
    5 3/4″ = Dia.
    (146.05 MM)

6. FN-5
    1/2″ = Dia.
    (12.7 MM)

7. RC-2
    2 1/4″ = Dia.
    (57.15 MM)

8. LC-10
    3 1/8″ = Dia.
    (79.375 MM)

Practice Exercise 7-6 and 7-7

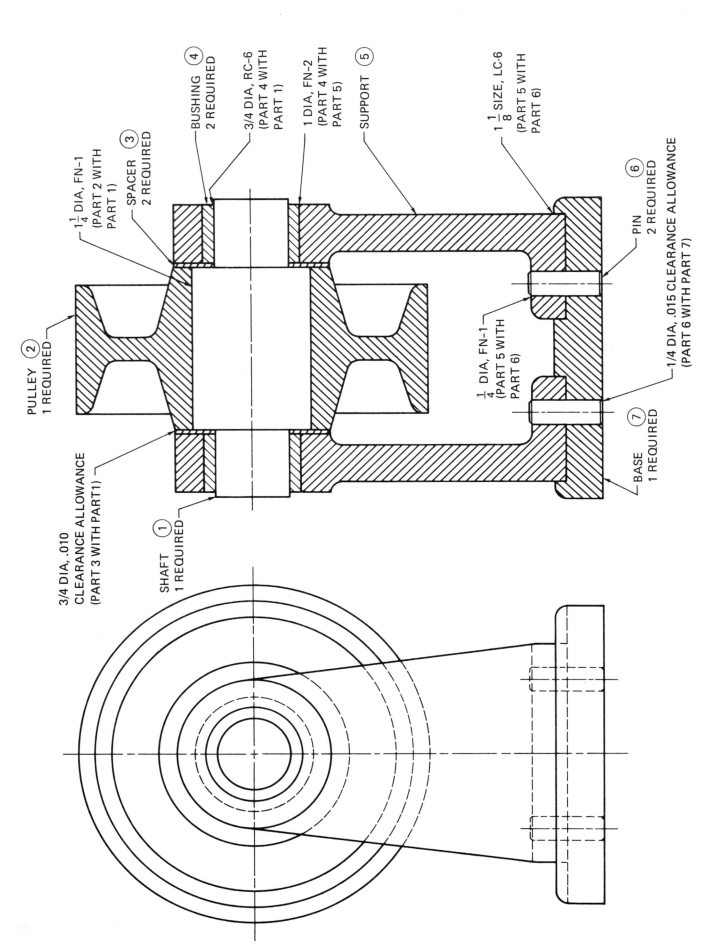

BUSHING ④
2 REQUIRED

3/4 DIA, RC-6
(PART 4 WITH
PART 1)

1 DIA, FN-2
(PART 4 WITH
PART 5)

SUPPORT ⑤

$1\frac{1}{8}$ SIZE, LC-6
(PART 5 WITH
PART 6)

$1\frac{1}{4}$ DIA, FN-1
(PART 2 WITH
PART 1)

SPACER ③
2 REQUIRED

PULLEY ②
1 REQUIRED

PIN ⑥
2 REQUIRED

1/4 DIA, .015 CLEARANCE ALLOWANCE
(PART 6 WITH PART 7)

$\frac{1}{4}$ DIA, FN-1
(PART 5 WITH
PART 6)

BASE ⑦
1 REQUIRED

3/4 DIA, .010
CLEARANCE ALLOWANCE
(PART 3 WITH PART1)

SHAFT ①
1 REQUIRED

PULLEY ASSEMBLY

Pages 158 through 161 show an assembly drawing and detail part drawings of a pulley. Pages 161 through 165 list questions about limits, tolerances, and allowances that pertain to these drawings. Tolerances to be used (unless otherwise noted) are:

$$.xx\ \ \ = \pm .015$$
$$.xxx\ \ = \pm .010$$
$$.xxxx = \pm .0002$$

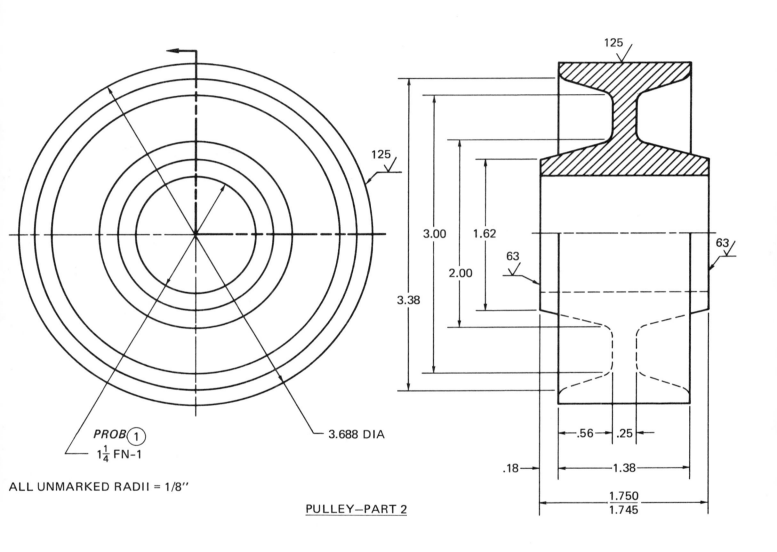

PROB(1)
1¼ FN-1

ALL UNMARKED RADII = 1/8″

3.688 DIA

PULLEY—PART 2

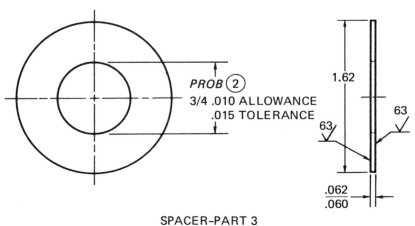

PROB (2)
3/4 .010 ALLOWANCE
.015 TOLERANCE

SPACER—PART 3

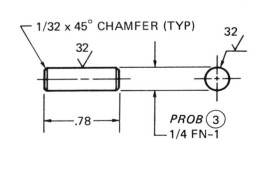

1/32 x 45° CHAMFER (TYP)

PROB (3)
1/4 FN-1

PIN—PART 6

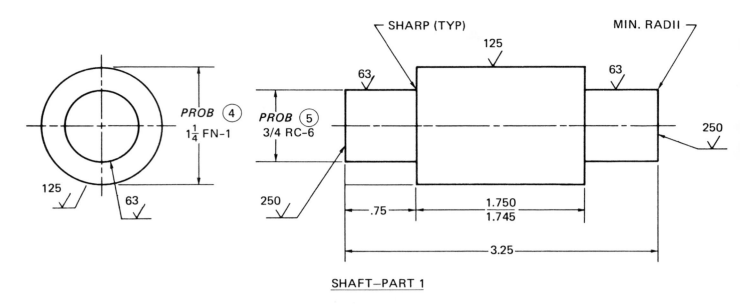

SHAFT—PART 1

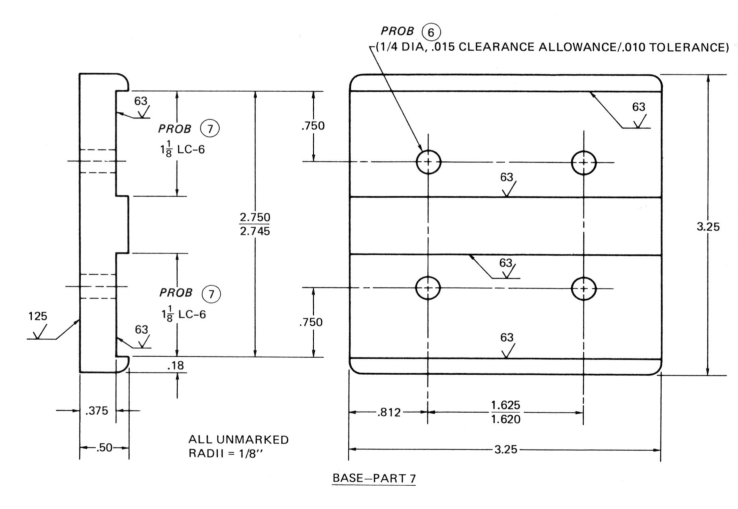

ALL UNMARKED
RADII = 1/8"

BASE—PART 7

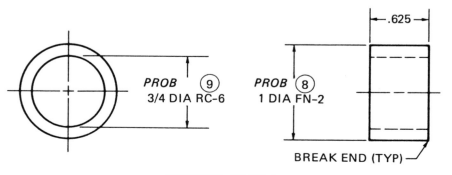

BREAK END (TYP)

BUSHING—PART 4

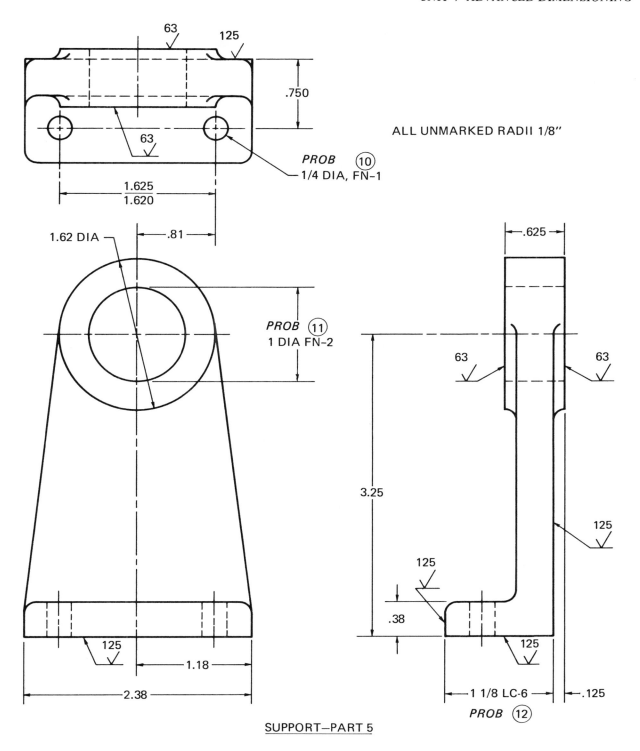

ALL UNMARKED RADII 1/8"

PROB ⑩
1/4 DIA, FN-1

PROB ⑪
1 DIA FN-2

PROB ⑫

SUPPORT—PART 5

Exercise 7-6

Place the answer in the space provided. Be sure to list limits in the correct order; i.e., smallest hole limit on top, largest shaft limit on top. Compare your work to the answers on pages 181 through 183.

1. What is the inside diameter of the pulley wheel (part 2)? This part fits on the shaft (part 1). 1 1/4-inch nominal size, FN-1.

2. What is the inside diameter of the spacer (part 3)? This part fits on the shaft (part 1). 3/4-inch nominal size, .010-inch clearance allowance, .015-inch tolerance.

3. What is the diameter of the pin (part 6)? This part fits tight on the support (part 5). 1/4-inch nominal size, FN-1.

4. What is the diameter of the largest shaft (part 1)? This part fits into the pulley (part 2). 1 1/4-inch nominal size, FN-1.

5. What is the diameter of the smallest shaft (part 1)? This part fits into the bushing (part 4). 3/4-inch nominal size, RC-6.

6. What is the diameter of the four holes in the base (part 7)? The four pins (part 6) fit into the base. 1/4-inch nominal size, .015-inch clearance allowance, .010-inch tolerance.

7. Using the same method used to calculate holes and shafts, figure the size of the slot (hole) in the base (part 7). The support (part 5) fits into this slot. Figure it as the key (shaft). 1 1/8-inch nominal size, LC-6.

8. What is the outside diameter of the bushing (part 4)? This part fits into the support (part 5). 1-inch nominal size, FN–2. Note that, in this case, the O.D. is actually considered the shaft.

9. What is the inside diameter of the bushing (part 4)? The shaft (part 1) fits into the bushing. 3/4-inch nominal size, RC-6.

10. What is the diameter of the four holes in the support (part 5)? The four pins (part 6) fit into the support. 1/4-inch nominal size, FN-1.

11. What is the diameter of the hole in the support (part 5)? The bushing (part 4) is pushed into this hole. 1-inch nominal size, FN-2.

12. Using the same method used to figure holes and shaft, figure the size of the key (shaft) in the support (part 5). This part fits into the base (part 7), which is considered the slot (hole). 1 1/8-inch nominal size, LC-6.

Exercise 7-7

Answer the questions in the space provided. Compare your work to the answers on page 183.

1. *Pulley (part 2).* What is the maximum diameter of the pulley?

2. *Pulley (part 2).* What is the minimum size from finish surface 63 to finish surface 63?

3. *Pin (part 6).* What is the maximum length of this part?

4. *Shaft (part 1).* If all dimensions are made to maximum limits, what will be the right-hand 3/4-inch diameter length?

5. *Base (part 7).* What is the minimum distance the 1/4-inch diameter holes can be apart and still be within limits?

6. *Base (part 7).* If all dimensions are made to minimum limits, what will be the right-hand dimension from the 1/4-inch diameter hole to the right-hand edge?

7. *Base (part 7).* What is the maximum depth of the two grooves?

8. *Support (part 5).* What are the limits from the bottom of the base to the center of the shaft?

9. *Support (part 5), Bushing (part 4).* If the support thickness at the top (1.62-inch diameter area) is made to maximum limit and the bushing length is made to minimum limit, what is the maximum depth to which the bushing could set into the support?

10. *Base (part 7), Support (part 5), and Pin (part 6).* If the .750-inch hole location in the base (part 7) was made to maximum limit, the pin (part 6) location in support (part 5) (.750-inch dimension) was made to minimum limit, and all parts at MMC (i.e. pin maximum size and hole minimum size), is it possible to put the parts together? If there is clearance, how much? If there is interference, how much? Explain in full.

TOLERANCING DRILLED HOLES

The chart in Appendix D is a guide for determining the limits for a hole's dimensions when a particular size drill is used. Steps one through four outline how to use the charts. The example uses a #45 standard drill.

Step 1. Locate the letter or number of the drill at the left of the chart.

Step 2. Convert it to a decimal:

$$\# 45 = .0820$$

Step 3. Find the upper limit:

$$.0820 + \text{tolerance of } .0043 = .0863$$

Step 4. Find the lower limit:

$$.0820 - \text{tolerance of } .0010 = .0810$$

The hole callout is: .0810 lower limit
                          .0863 upper limit

The term "drill" is not used.

Practice Exercise 7-8

Find the limits for the following size drilled holes using the drilled hole tolerance charts in Appendix D. Compare your work to the answers on page 184.

1. 1/2-size drill

2. 23/64-size drill

3. 1/8-size drill

4. X-size drill

SURFACE TEXTURE

Surface texture that appears smooth to the eye may be very rough. When viewed under optical devices, it is a complex structure. However, for the purpose of this unit, the most commonly encountered surface control principles are discussed.

The most common surface irregularity is measured from the highest point to the lowest point and recorded in microinches (one millionth of an inch, figure 7-25).

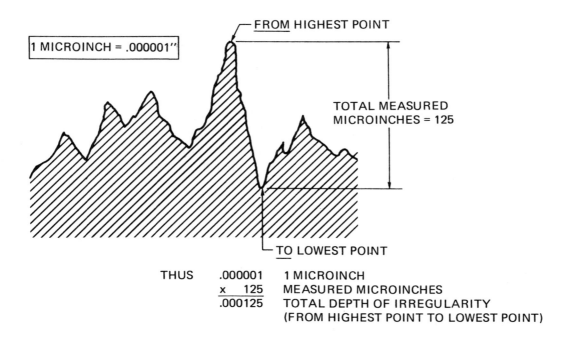

```
                                     FROM HIGHEST POINT

1 MICROINCH = .000001"
                                                  TOTAL MEASURED
                                                  MICROINCHES = 125



                                     TO LOWEST POINT

          THUS    .000001   1 MICROINCH
                 x    125   MEASURED MICROINCHES
                  .000125   TOTAL DEPTH OF IRREGULARITY
                           (FROM HIGHEST POINT TO LOWEST POINT)
```

Fig. 7-25 Surface irregularity

Surface irregularity can be measured using a microfinish comparator, figure 7-26.

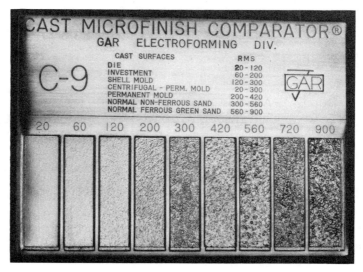

Fig. 7-26 Microfinish comparator

Surface texture symbols can be modified to indicate special designations. For instance, a horizontal bar added to the symbol, figure 7-27 (A) means that material must be removed by machining to produce the surface texture desired. Add a number to the left of that to indicate the amount of material to be removed, (B). A circle placed within the vee means that the surface must be machined without removing any material, (C). When information other than roughness must be included, a horizontal extension is added to the symbol, (D).

Fig. 7-27 Surface texture symbols

Finishes desired:
- Sand castings — 1000 to 300 microinches
- Forgings — 500 to 100 microinches
- Die castings — 150 to 16 microinches
- Extruded shapes — 150 to 10 microinches
- Drilled holes — 160 to 80 microinches
- Reamed holes — 140 to 50 microinches
- Broached holes — 125 to 40 microinches

Do not place finish marks on drilled, reamed, or broached holes unless a smoother surface than indicated above is required.

GEOMETRIC TOLERANCING

Up to this point, permissible variation of size has been used. *Geometric tolerancing* is the permissible variation of shape. Most shapes are broken down into a basic geometric form, such as a plane (surface), cylinder, cone, square, hex, etc. Nothing can be made exactly the same size or shape everytime, so a national system of tolerancing was developed. For size, a system of limits and tolerances are used. For shape, geometric tolerancing is used. Geometric tolerancing is used only where the shape could be critical to the function of the part or could effect the interchangeablility of parts.

Some companies do not use geometric tolerancing while others use it extensively. The next few illustrations give a brief idea of what geometric tolerancing is and how it works. Geometric tolerancing is applied to:

- Flatness
- Straightness
- Cylindricity
- Roundness
- Parallelism
- Perpendicularity
- Angularity
- Concentricity
- Symmetry

Feature Control Symbol

Feature control symbols call out what planes (views) the tolerances will be in respect to. Sometimes a tolerance is tied to more than one plane. In this case the first, second, and auxiliary datum notes are used, figure 7-28.

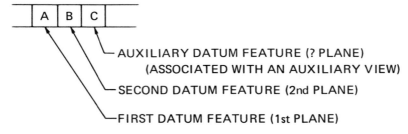

Fig. 7-28 Feature control symbol

If any surface does not have a geometric tolerance specified, the form is allowed to vary within the given limits of size.

Flatness ⟋▱⟋

Flatness means that the entire surface must lie between two parallel planes that cannot be more than a specified tolerance apart. The symbol for flatness is a *parallelogram*, figure 7-29. Note that flatness applies to flat surfaces only.

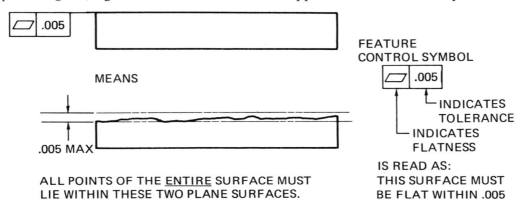

Fig. 7-29 Geometric tolerancing, flatness

Straightness (−)

Straightness means that the entire surface must be straight within a given limit. Straightness applies to cylinder or cone surfaces only, figure 7-30.

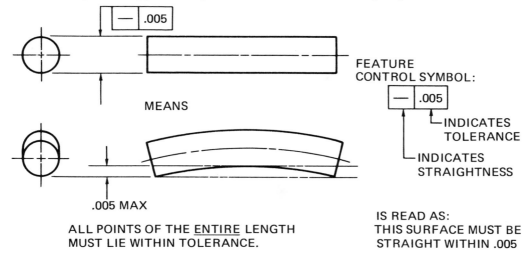

Fig. 7-30 Geometric tolerancing, straightness

Cylindricity

Cylindricity is a combination of straightness, roundness, and parallelism. A cylindricity tolerance refers to all three of these properties. Note that both circles, figure 7-31, are about the same axis.

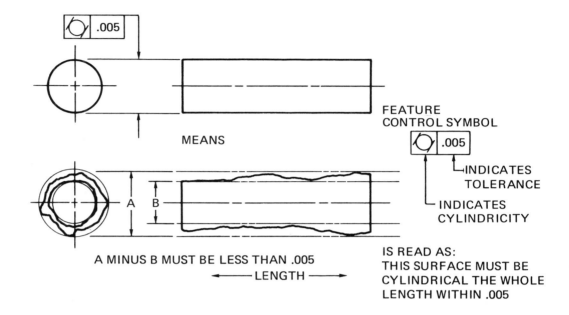

MEANS

FEATURE CONTROL SYMBOL

INDICATES TOLERANCE

INDICATES CYLINDRICITY

A MINUS B MUST BE LESS THAN .005

← LENGTH →

IS READ AS:
THIS SURFACE MUST BE CYLINDRICAL THE WHOLE LENGTH WITHIN .005

Fig. 7-31 Geometric tolerancing, cylindricity

Roundness

Roundness refers to the circular form of an object. A roundness tolerance describes the range within which the radius of a circle may vary. One of the features of a cylindrical object is that it has a round cross section, figure 7-32.

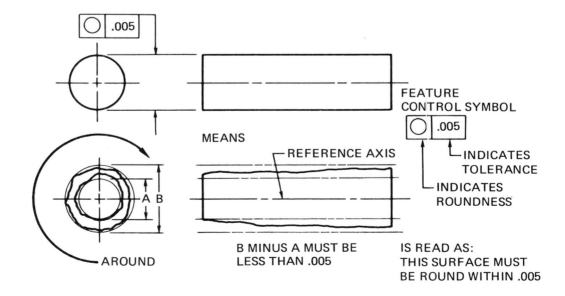

MEANS

FEATURE CONTROL SYMBOL

REFERENCE AXIS

INDICATES TOLERANCE

INDICATES ROUNDNESS

AROUND

B MINUS A MUST BE LESS THAN .005

IS READ AS:
THIS SURFACE MUST BE ROUND WITHIN .005

Fig. 7-32 Geometric tolerancing, roundness

Parallelism ‖

Parallelism means that an entire surface must lie between two parallel planes that cannot be more than a specified tolerance apart, parallel to and in relation to a given surface or datum. In figure 7-33, the bottom surface is datum "A."

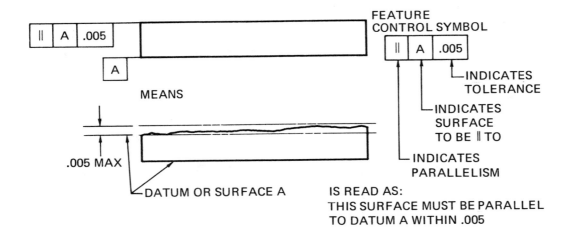

Fig. 7-33 Geometric tolerancing, parallelism

Remember a datum could be a surface, point, center line or any feature on the object.

Perpendicularity ⊥

Perpendicularity means that an entire surface must lie between two parallel planes that cannot be more than a specified tolerance apart, perpendicular to and in relation to a given surface or datum. The upright surface in figure 7-34 is datum "A."

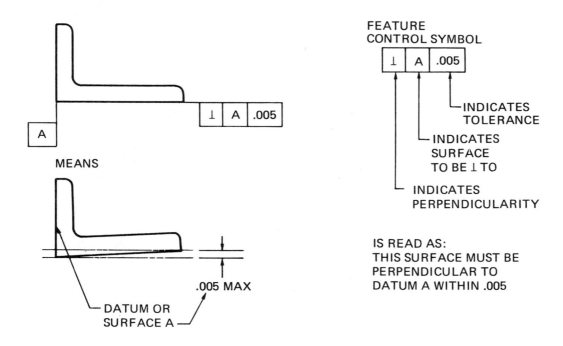

Fig. 7-34 Geometric tolerancing, perpendicularity

Angularity ∠

Angularity means that the entire surface must lie between two parallel planes which are at the true angle in relation to a specified surface or datum (more than one datum can be used) with a specified tolerance, figure 7-35.

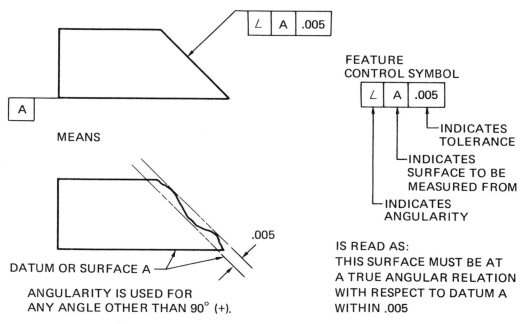

Fig. 7-35 Geometric tolerancing, angularity

Concentricity ◎

Concentricity means that the axis of one feature or diameter must lie within a cylindrical tolerance zone, which is concentric to the datum axis of another feature or diameter within a specified tolerance, figure 7-36.

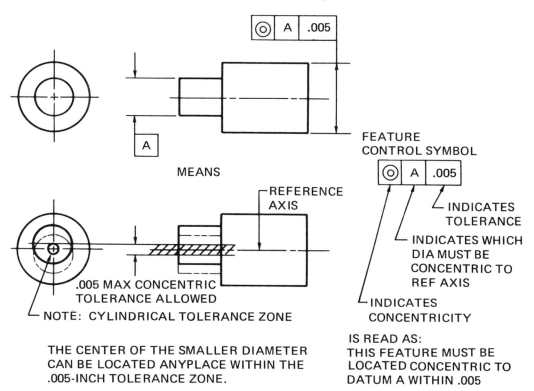

Fig. 7-36 Geometric tolerancing, concentricity

Symmetry ⚌

Symmetry means that the entire feature must lie between two parallel planes that cannot be more than a specified tolerance apart symmetrically and in regards to a surface(s) or datum. The specified tolerance in this case must be *equally spaced between the datum,* or surface(s), figure 7-37.

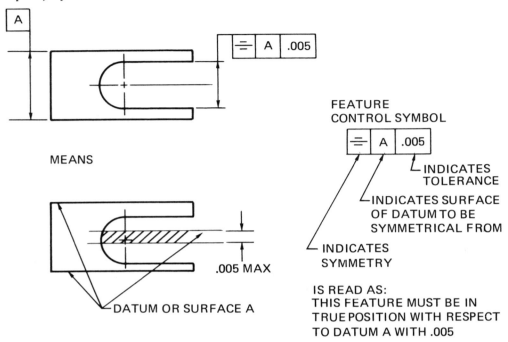

Fig. 7-37 Geometric tolerancing, symmetry

Modifiers

Sometimes a *modifier* is added to the feature control symbol:

- Ⓢ means "regardless of feature size" (rfs). This reads as: parallel to surface A, regardless of feature size and within .005 inch.
- Ⓜ means "maximum material condition" (MMC). This reads as: parallel to surface A when it is at maximum material condition only and within .005 inch.

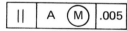

Total Indicator Reading (TIR)

The note TIR on a drawing means *total indicator reading.* This means the part to be checked must be round the full 360 degrees to within .005 inch. The part is set up so it can rotate about a fixed center line. An indicator is mounted above it and set on 0. The part is then rotated and the indicator must not move more than a total of .005 inch in both directions. If it does, it is not within tolerance, figure 7-38.

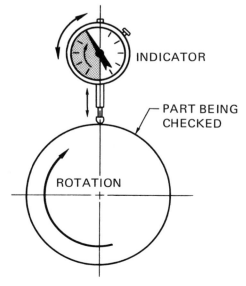

Fig. 7-38 Dial indicator measuring tolerance

Practice Exercise 7-9

Carefully study each control note on the sample drawing (1 through 8). In the space provided, explain in full what each control note means. Compare your work to the answers on page 184.

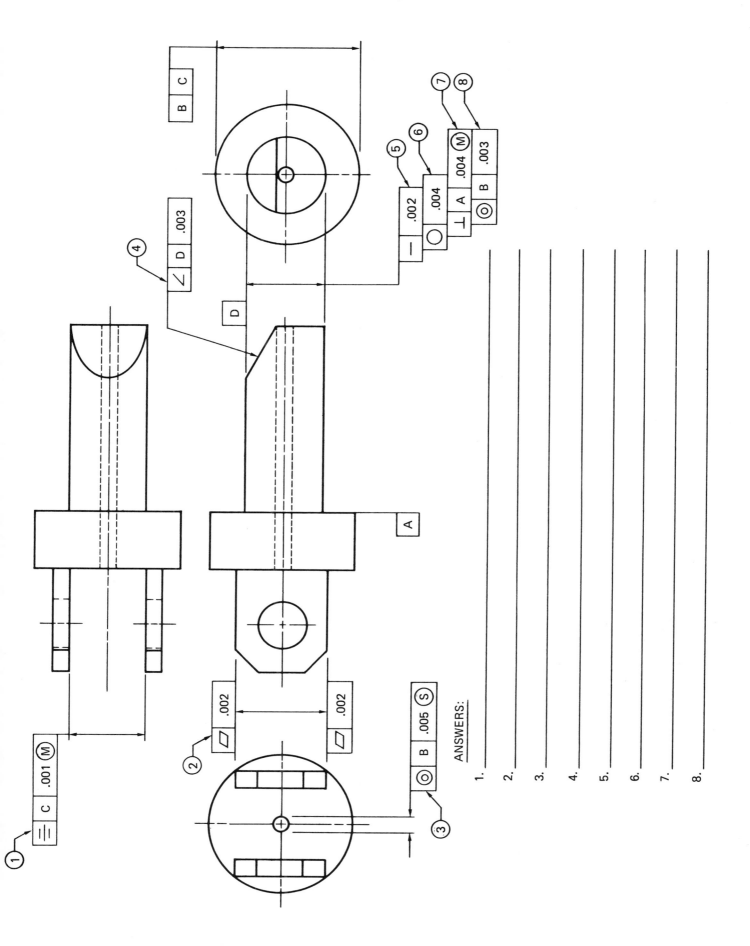

ANSWERS:

1. _____

2. _____

3. _____

4. _____

5. _____

6. _____

7. _____

8. _____

TRUE POSITION

Using the normal bilateral system of dimensioning, a hole with limits of + or – .015-inch is dimensioned as in figure 7-39.

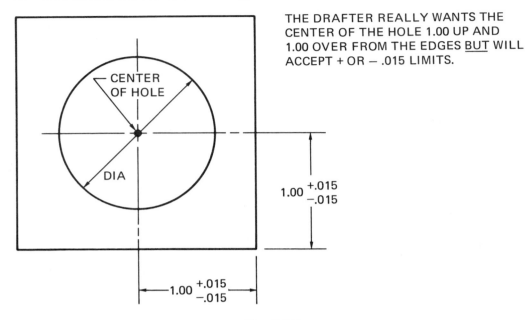

THE DRAFTER REALLY WANTS THE CENTER OF THE HOLE 1.00 UP AND 1.00 OVER FROM THE EDGES <u>BUT</u> WILL ACCEPT + OR – .015 LIMITS.

Fig. 7-39

If the above limits are accepted, the *tolerance zone* forms a square area that, in effect, establishes boundaries in which the center of the circle is located. In figure 7-40, the .030-inch tolerance zone extends the full length of the hole.

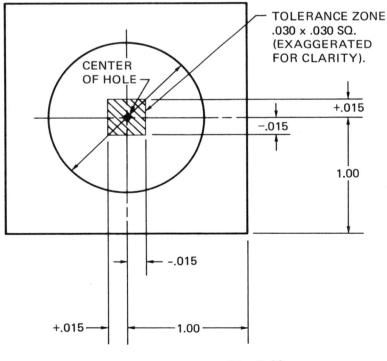

TOLERANCE ZONE .030 x .030 SQ. (EXAGGERATED FOR CLARITY).

Fig. 7-40

Example 1. The four examples in figure 7-41, illustrate the worst conditions, maximum limits. Vertically and horizontally, the center of the hole can be from the design size and still be within limits.

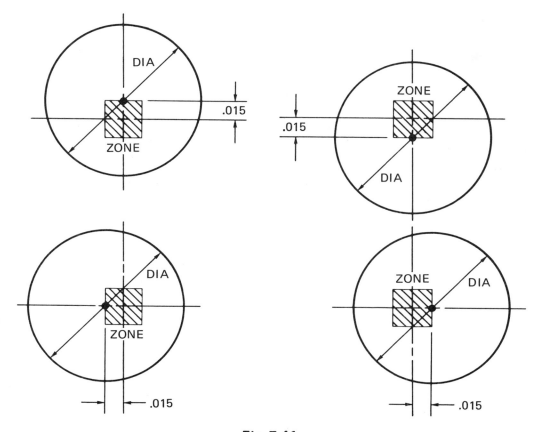

Fig. 7-41

Example 2. The four examples in figure 7-42 show the worst conditions, maximum limits. Diagonally across corners, the center of the hole can be from the design size and still be within limits.

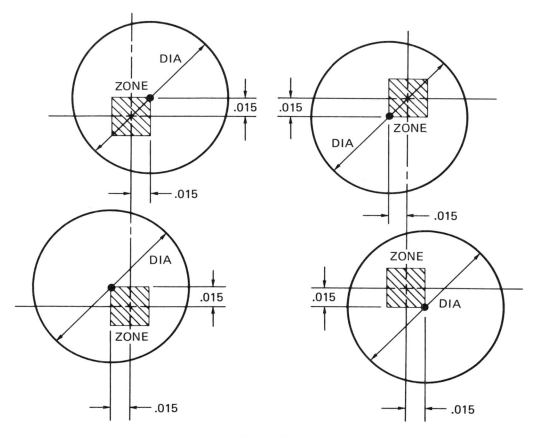

Fig. 7-42

In example 1, figure 7-41, the maximum distance from the design size still within limits is .015 inch, figure 7-43.

In example 2, figure 7-42, the maximum distance from the design size is much more, figure 7-44.

Using simple trigonometry, it is possible to calculate the exact diagonal distance from the center of the hole (.021-inch radii). In effect, there are two sizes or limits used in locating the center of the hole, ± .015 inch one way and ± .021 inch the other way, using *bilateral tolerancing,* figure 7-45.

If .021 inch is within tolerance, allow a .021-inch radius all around. A circle has 57 percent more area than a square. This will reduce scrap, reduce inspection time, reduce cost, and, in effect, still allow the exact same limits as the bilateral system (across corners), figure 7-46.

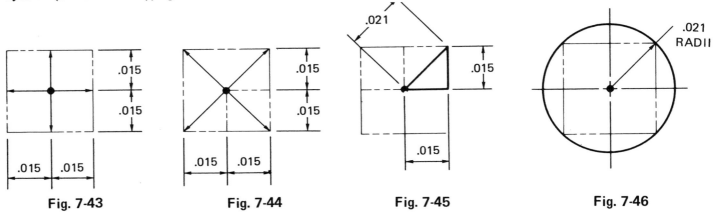

| Fig. 7-43 | Fig. 7-44 | Fig. 7-45 | Fig. 7-46 |

Simply stated, *true position* (TP) *enlarges* the tolerance zone. The symbol for true position is: ⊕

Comparing the bilateral tolerancing system with the true positioning system, it is easy to see the tolerance zone using true positioning is much larger, figure 7-47. This will result in fewer rejections of parts and lower costs. There is a 57% larger tolerance zone using true positioning.

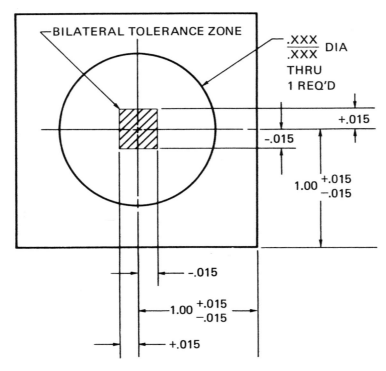

Fig. 7-47

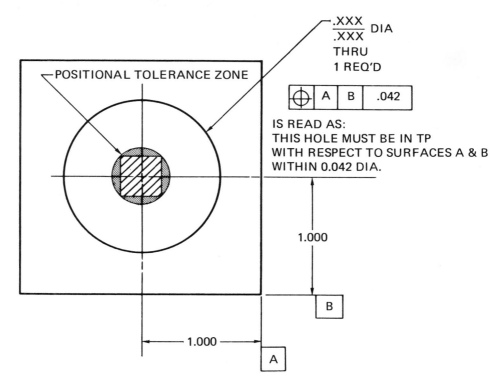

$\dfrac{.XXX}{.XXX}$ DIA
THRU
1 REQ'D

| ⊕ | A | B | .042 |

IS READ AS:
THIS HOLE MUST BE IN TP
WITH RESPECT TO SURFACES A & B
WITHIN 0.042 DIA.

POSITIONAL TOLERANCE ZONE

1.000

B

1.000

A

Fig. 7-47 (Continued)

Figure 7-48 is a list of most of the standard symbols used in engineering.

SYMBOLS

Symbol	Name
▱	FLATNESS
—	STRAIGHTNESS
⌭	CYLINDRICITY
◯	ROUNDNESS
‖	PARALLELISM
⊥	PERPENDICULARITY
∠	ANGULARITY
◎	CONCENTRICITY
⩳	SYMMETRY
Ⓢ	REGARDLESS OF FEATURE SIZE
Ⓜ	MAXIMUM MATERIAL CONDITION
⊕	TRUE POSITION

Fig. 7-48 Symbol Chart

ANSWERS TO PRACTICE EXERCISES

Carefully compare your work to the answer for each exercise. Refer any questions to the instructor.

Exercise 7-1

Ⓐ 1. _____.500_____ MMC of hole

2. _____.498_____ MMC of shaft

3. ___+ .002___ Minimum clearance

4. ___+ .006___ Maximum clearance

5. __Clearance__ Kind of fit

Minimum Allowance	Maximum Allowance
.500 H	.502 H
− .498 S	− .496 S
+ .002 clearance	+ .006 clearance

Parts always clear each other.

Ⓑ 1. _____.494_____ MMC of hole

2. _____.501_____ MMC of shaft

3. ___− .007___ Maximum interference

4. ___− .004___ Minimum interference

5. __Interference__ Kind of fit

.494 H	.496 H
− .501 S	− .500 S
− .007 interference	− .004 interference

Parts always interfere with each other.

Ⓒ 1. _____.500_____ MMC of hole

2. _____.507_____ MMC of shaft

3. ___− .007___ Maximum interference

4. ___+ .003___ Maximum clearance

5. __Transition__ Kind of fit

.500 H	.505 H
− .507 S	− .502 S
− .007 interference	+ .003 clearance

Parts sometimes clear and sometimes interfere.

Exercise 7-2

Prob.	Nom. Size	Basic Size	Dim. of Hole	Dim. of Shaft	Hole Tol.	Shaft Tol.	Min. Allowance	Max. Allowance
1	5/8	.625	.625 / .630	.623 / .618	.005	.005	+ .002	+ .012
2	1 1/4	1.250	1.250 / 1.251	1.249 / 1.247	.001	.002	+ .001	+ .004
3	5/16	.312	.312 / .314	.316 / .314	.002	.002	− .004	− .000
4	3/4	.750	.750 / .753	.749 / .747	.003	.002	+ .001	+ .006
5	13/16	.812	.812 / .813	.815 / .814	.001	.001	− .003	− .001

①. .625 = 5/8

.625 H
− .002 Min. Allow.
.623 S
− .005 Tol. (S)
.618 S

.630 H
− .625 H
.005 Tol. (H)

.630 H
− .618 S
+ .012 Max. Allow.

②. 1 1/4 = 1.250

1.250 H
− .001 Min. Allow.
1.249 S
− .002 Tol (S)
1.247 S

1.250 H
+ .001 Tol. (H)
1.251 H

1.251 H
− 1.247 S
+ .004 Max. Allow.

(3.) 3.12 = 5/16   (4.) 3/4 = .750

.314 H   .316 S   .750 H   .750 H
– .312 H   – .314 S   + .003 Tol. (H) – .001 Min. Clear.
.002 Tol. (H)   .002 Tol. (S)   .753 (H)   .749 (S)

.312 H   .314 H   .749 S   .753 H
– .316 S   – .314 S   – .002 Tol.   – .747 S
– .004 Max.   – .000 Min.   .747 (S)   + .006 Max.
   Inter.   Inter.   Clear.

(5.) .812 = 13/16

.813 H   .815 S
– .812 H   – .814 S
.001 Tol. (H)   .001 Tol. (S)

.812 H   .813
– .815 S   – .814
– .003 Max. Inter. – .001 Min. Inter.

Exercise 7-3

			Min. Allowance	Max. Allowance
(A)	1.	.625 MMC of slot	.625 S	.627 S
			– .623 K	– .621 K
	2.	.623 MMC of key	+ .002 Clearance	+ .006 Clearance
	3.	+ .002 Min. clearance		
	4.	+ .006 Max. clearance	Parts always clear each other.	
	5.	Clearance Kind of fit		
(B)	1.	.744 MMC of slot	.744 S	.746 S
			– .751 K	– .750 K
	2.	.751 MMC of key	– .007 Interference	– .004 Interference
	3.	– .007 Max. interference		
	4.	– .004 Min. interference	Parts always interfere with each other.	
	5.	Interference Kind of fit		
(C)	1.	.375 MMC of slot	.375 S	.380 S
			– .382 K	– .377 K
	2.	.382 MMC of key	– .007 Interference	+ .003 Clearance
	3.	– .007 Max. interference		
	4.	+ .003 Max. clearance	Parts sometimes clear/sometimes interfere.	
	5.	Transition Kind of fit		

Exercise 7-4

1. The groove is the most important feature, the limits are the closest. Next would come the location of the two holes and the size of the largest hole.
2. 63 microinches
3. The .50-inch diameter holes would probably be drilled because there is only a two-place decimal, indicating ± .015, which is very loose tolerance.

4. The .7500 diameter hole would probably be reamed because there is a four-place decimal, indicating ± .0002, which is a very close tolerance.

5. It is a finished surface, 125 microinches, and is larger than the top surface.

6. .7500 + .0002 = .7502 upper limit
    .7500 = .0002 = .7498 lower limit

7. With the 2.0000 dimension at the minimum limit (1.9998) and the 1.0000 dimension at the maximum limit (1.0002), the closest the two holes can be and still be within limits is 1.9998 minus 1.0002 equals .9996 minimum dimension.

8. Unilateral dimension (one-direction). Note: the groove is exactly like a hole, thus, the smallest limit must go on top.

9. 2.2500 + .0002 = 2.2502 maximum height; 2.2500 minus .0002 = 2.2498 minimum height. Maximum tolerance = .0004.

10. The groove height is 2.1251/2.1252. Note: the groove is exactly like a hole, thus, the smallest limit must go on top.

11. Two-placed decimals are used anyplace a size is not important to the looks or function of the part to reduce time and cost to manufacture the part.

12. .6250 + .0002 = .6252 + 2.1252 = 2.7502 maximum height allowed.

Exercise 7-5

① RC-8
1 1/2 Dia.

Metric	Inch	
38.100	1.5000	smallest hole (basic size)
+ .100	+ .0040	hole tolerance
38.200	1.5040	largest hole
38.100	1.5000	basic size
− .130	− .0050	allowance
37.970	1.4950	largest shaft
37.970	1.4950	largest shaft
− .060	− .0025	shaft tolerance
37.910	1.4925	smallest shaft

② LC-2
5/8 Dia.

Metric	Inch	
15.875	.6250	smallest hole (basic size)
+ .018	+ .0007	hole tolerance
15.893	.6257	largest hole
15.875	.6250	basic size
− .000	− .0000	allowance
15.875	.6250	largest shaft
15.875	.6250	largest shaft
− .010	− .0004	shaft tolerance
15.865	.6246	smallest shaft

③ LN-3
3.0 Dia.

Metric	Inch	
76.200	3.0000	smallest hole (basic size)
+ .030	+ .0012	hole tolerance
76.530	3.0012	largest hole
76.200	3.0000	basic size
+ .059	+ .0023	allowance
76.259	3.0023	largest shaft
76.259	3.0023	largest shaft
− .018	− .0007	shaft tolerance
76.241	3.0016	smallest shaft

④ RC-3
1 7/16 Dia.

Metric	Inch	
36.5125	1.4375	smallest hole (basic size)
+ .0300	+ .0010	hole tolerance
36.5425	1.4385	largest hole
36.5125	1.4375	basic size
− .0300	− .0010	allowance
36.4825	1.4365	largest shaft
36.4825	1.4365	largest shaft
− .0150	− .0006	shaft tolerance
36.4675	1.4359	smallest shaft

5. LT–4
   5 3/4 Dia.

Metric	Inch	
146.050	5.7500	smallest hole (basic size)
+ .064	+ .0025	hole tolerance
146.114	5.7525	largest hole
146.050	5.7500	basic size
+ .044	+ .0017	allowance
146.094	5.7517	largest shaft
146.094	5.7517	largest shaft
− .041	− .0016	shaft tolerance
146.053	5.7501	smallest shaft

6. FN–5
   1/2 Dia.

Metric	Inch	
12.700	.5000	smallest hole (basic size)
+ .025	+ .0010	hole tolerance
12.725	.5010	largest hole
12.700	.5000	basic size
+ .058	+ .0023	allowance
12.758	.5023	largest shaft
12.758	.5023	largest shaft
− .018	− .0007	shaft tolerance
12.740	.5016	smallest shaft

7. RC–2
   2 1/4 Dia.

Metric	Inch	
57.150	2.2500	smallest hole (basic size)
+ .018	+ .0007	hole tolerance
57.168	2.2507	largest hole
57.150	2.2500	basic size
− .010	− .0004	allowance
57.140	2.2496	largest shaft
57.140	2.2496	largest shaft
− .013	− .0005	shaft tolerance
57.127	2.2491	smallest shaft

8. LC–10
   3 1/8 Dia.

Metric	Inch	
79.375	3.125	smallest hole (basic size)
+ .310	+ .012	hole tolerance
79.685	3.137	largest hole
79.375	3.125	basic size
− .250	− .010	allowance
79.125	3.115	largest shaft
79.125	3.115	largest shaft
− .180	− .007	shaft tolerance
78.945	3.108	smallest shaft

Exercise 7-6

1. 1 1/4
   FN–1

   1.2500 smallest hole (basic size)
   + .0006 hole tolerance
   1.2506 largest hole

   1.2500 basic size
   + .0013 allowance
   1.2513 largest shaft

   1.2513 largest shaft
   − .0004 shaft tolerance
   1.2509 smallest shaft

   Answer Prob. 1   1.2500
                    1.2506

   Answer Prob. 4   1.2513 = longest shaft dia.
                    1.2509

2. 3/4
   .010 Allow./.015 Tol.

   Note: Prob. 5 must be completed in order to complete this problem. . .

   .7484 largest shaft
   + .0100 allowance
   .7584 smallest hole

   .7584 smallest hole
   + .0150 tolerance
   .7734 largest hole

   Answer Prob. 2   .7584
                    .7734

   If this dimension is not important, or if the fit is too close to any mating part, a diameter of .758/.773 would be better.

3. 1/4
   FN–1

     .2500 smallest hole (basic size)
   + .0004 hole tolerance
     .2504 largest hole

     .25000 basic size
   + .00075 allowance
     .25075 largest shaft

     .25075 largest shaft
   – .00025 shaft tolerance
     .25050 smallest shaft

   Answer Prob. 3 $\dfrac{.2508}{.2505}$

   Answer Prob. 10 $\dfrac{.25075}{.25050}$ or $\dfrac{.2508}{.2505}$

4. 1 1/4   (See Answer Prob. 1)
   FN–1

5. 3/4
   RC–6

     .7500 smallest hole (basic size)
   + .0020 hole tolerance
     .7520 largest hole

     .7500 basic size
   – .0016 allowance
     .7484 largest shaft

     .7484 largest shaft
   – .0012 shaft tolerance
     .7472 smallest shaft

   Answer Prob. 9 $\dfrac{.7500}{.7520}$

   Answer Prob. 5 $\dfrac{.7484}{.7472}$

6. 1/4
   .015 Allow./.010 Tol.

   Note: Prob. 3 must be completed in
         order to complete this problem.

     .25075 largest shaft
   + .01500 allowance
     .26575 smallest hole

     .26575 smallest hole
   + .01000 tolerance
     .27575 largest hole

   Answer Prob. 6 $\dfrac{.26575}{.27575}$

   If this dimension is not important, or if
   the fit is too close to any mating parts, a
   diameter of .266/.276 would be better.

7. 1 1/8
   LC–6

     1.1250 smallest slot (basic size)
   + .0020 hole tolerance
     1.1270 largest slot

     1.1250 basic size
   – .0008 allowance
     1.1242 largest key

     1.1240 largest key
   – .0012 shaft tolerance
     1.1230 smallest key

   Answer Prob. 7 $\dfrac{1.1250}{1.1270}$ slot

   Answer Prob. 12 $\dfrac{1.1242}{1.1230}$ key

8. 1
   FN–2

     1.0000 smallest hole (basic size)
   + .0008 hole tolerance
     1.0008 largest hole

     1.0000 basic size
   + .0019 allowance
     1.0019 largest shaft

     1.0019 largest shaft
   – .0005 shaft tolerance
     1.0014 smallest shaft

   Answer Prob. 11 $\dfrac{1.0000}{1.0008}$

   Answer Prob. 8 $\dfrac{1.0019}{1.0014}$

9. 3/4
RC-6
(See Answer Prob. 5)

11. 1
FN-2
(See Answer Prob. 8)

10. 1/4
FN-1
(See Answer Prob. 3)

12. 1 1/8
LC-6
(See Answer Prob. 7)

To help eliminate simple errors, it is a good idea to do all math problems neatly, in some kind of order, and with each figure labeled as illustrated above. Work out a "system" that is best for you and get in the habit of *always* using that system. Engineering dimensions *must* be 100% correct. Check and re-check all dimensions. Stop and review the detail drawings used in this assignment. Note the various methods used to dimension each detail.

Exercise 7-7

1. $3.688 + 0.010 = 3.698$ maximum diameter.
2. Minimum size allowance 1.745
3. $.78 + .015 = .795$ maximum length.
4. $.75 + .015 = .765 + 1.750 = 2.515$
   $3.25 + .015 = 3.265 - 2.515 = .750$
5. 2.745 minimum limit/$.750 + .010 = .760$ maximum $\times 2 = 1.520$
   2.745 minus (-) $1.520 = 1.225$ minimum distance between the 1/4 diameter holes.
6. $.812 - .010 = .802$ minimum dimension $+ 1.620$ minimum dimension. $= 2.422/3.25 - .015 = 3.235$ minimum dimension$/3.235 - 2.422 = .813$.
7. $.50 + .015 = .515$ max. height$/.375 - .010 = .365$ min. size$/.515 - .365 = .150$ maximum depth allowed.
8. $3.25 + .015 = 3.265$ maximum limit of height.
   $3.25 - .015 = 3.235$ minimum limit of groove.
9. $.625 + .010 = .635$
   $.625 - .010 = .615/.635$ maximum width support $- .625$ minimum width bushing $= .020$ total depth bushing *could* be set in and still be within limits.
10. Using same base line:
    Base $.750 + .010$ Tol. $= .760$ maximum distance from base line.
    $.760 - .1328$ (1/2 *minimum* hole size) $= .6272$ from base line to right side edge of hole in base.
    Support $.750 - .010$ Tol. $= .740$ minimum distance from base line.
    $.740 - .1252$ (1/2 *maximum* pin size) $= .6148$ from base line to right side edge of pin in support. Thus, there would be *interference* and under some conditions the parts would *not* go together $.6272 - .6148 = .0124$ interference. This is a poor design, the holes in base, parts should have a larger allowance.
    (This can be proven also by drawing all parts per limits *10 times size,* using a common base line.)

Exercise 7-8

It cannot be stressed enough how important the size and location dimensions are. A drafter must be 100 percent correct on each and every dimension, every time. Double check each dimension before and after the drawing is completed. Do not call out the process such as "drill," "ream," "bore," etc. The smallest limit is placed on top for hole limit dimensions.

(1.) *1/2 Drill*

.5000	.5000			
+ .0079	− .0020	=	.4980	Dia. Hole
.5079	.4980		.5079	
Max.	Min.			

(2.) *23/64 Drill*

.3594	.3594			
+ .0071	− .0020	=	.3574	Dia. Hole
.3665	.3574		.3665	
Max.	Min.			

(3.) *1/8 Drill*

.1250	.1250			
+ .0050	− .0010	=	.1240	Dia. Hole
.1300	.1240		.1300	

(4.) *'X' Drill*

.3970	.3970			
+ .0073	− .0020	=	.3950	Dia. Hole
.4043	.3950		.4043	

Exercise 7-9

1. The two 'ears' must be in true position with respect to surface "C" within .001 when they are at maximum material condition.
2. This surface must be flat the whole length within .002.
3. The hole must be located concentric to surface "B" within .005 regardless of size.
4. This angle must be at the true specified angle with respect to surface "D" within .003.
5. This diameter must be straight within its full length within .005.
6. This diameter must be round the entire length within .004.
7. This diameter must be perpendicular to surface "A" within .004 at maximum material condition.
8. This diameter must be concentric to surface "B" within .003.

UNIT REVIEW

No time limit

1. What size hole can be expected from a 23/64 size drill? Show all math.

2. A drafter is designing a hole and a shaft for high speed under heavy pressure. What type of fit should be used?

3. What factors must be considered in deciding the tolerances placed on a drawing? List at least six items.

4. At what angle is a standard countersink screw head?

5. Using normal tolerances, is a 1/2-inch diameter hole the same size as a .500-inch diameter hole? Explain in full which hole costs the most to make.

6. Limits for a key and slot are figured the same as limits for what?

7. Give the limits for a hole and shaft with a 1 1/4-inch nominal size, .001-inch hole tolerance, .002-inch shaft tolerance, and + .001-inch allowance. What is the clearance?

8. What does the modifier "S" mean in regards to geometric tolerancing?

9. What is the symbol for true position?

10. What are the limits for a key and slot with a basic shape of 1.5685-inches using an interference allowance of .0006 inch, a tolerance of .0005 inch for the key, and a tolerance of .0003 inch for the slot?

Before proceeding to the next unit:

_____ Instructor's approval

_____ Progress plotted

UNIT 8

ASSEMBLY & DETAIL DRAWING

OBJECTIVE

The student will perform actual mechanical drafter's work by doing the design layouts, computations, and dimensioning of details, subassembly, and assembly drawings.

PRETEST

Two-hour time limit

Using the detail drawings on pages 190 through 198 of a model airplane engine, draw a complete assembly drawing. Use as many views and/or kinds of view(s) needed to fully illustrate each part and where they are located in the assembly. Use all standard drafting practices. Use balloon callouts for each part.

RELATED TERMS

Give a brief definition of each term as progress is made through this unit.

Assembly drawing _____

Design layout _____

Detail drawing _____

Purchased parts _____

KINDS OF DRAWINGS

There are various kinds of drawings used in a mechanical engineering department. The drafter must be able to recognize and draw each of them.

Design Layout Drawing

All major designs start from a design layout. This is usually a sketch drawing, full size or to scale. It is up to the drafter to draw each part depicted on this layout. In drawing the parts, the drafter must use the basic ideas in the design layout, but may change it to fit standard material stock, standard methods of manufacturing, and standard material sizes. If changes are made, they should be reviewed with the designer for his approval. Most of the time the designer draws parts as close to size as possible, but a design layout is not usually dimensioned unless particular dimensions must be maintained.

Assembly Drawing

Any product that has more than one part must have an assembly drawing. The assembly drawing shows how a product is assembled when completed. It can have one, two, three, or more views that are placed in the usual positions. One view is often a section view to illustrate the various parts and how they are assembled. Each part in an assembly drawing is identified by a circled detail number.

Permanently Fastened Parts Drawing

When two or more parts are permanently fastened such that they cannot be disassembled after assembly, they are called out as in a **subassembly drawing**.

Detail Drawing

Each part must have its own fully-dimensioned detail drawing, its own drawing number, and its own drawing title. All the information needed to manufacture the part is included in the detail drawing. The shape must be shown in the views given, features must be dimensioned and located, and specifications given on the drawing or title block.

Purchased Parts

A manufacturing company cannot afford to make standard items such as nuts, bolts, and washers which can be purchased ready for use. A drafter should try to design around standard parts whenever possible. Such parts are not drawn but simply called out on an assembly drawing by size, material, and finish. Other nonhardware parts may be treated in a similar fashion.

Practice Exercise 8-1

Using the design layout, make a detail drawing of each part with a circled part number. Use the tolerance chart in the lower right corner of the drawing. Thread callouts should be included on the drawings. Compare your work to the answers on pages 204 through 207. This drawing is dimensioned in inches only.

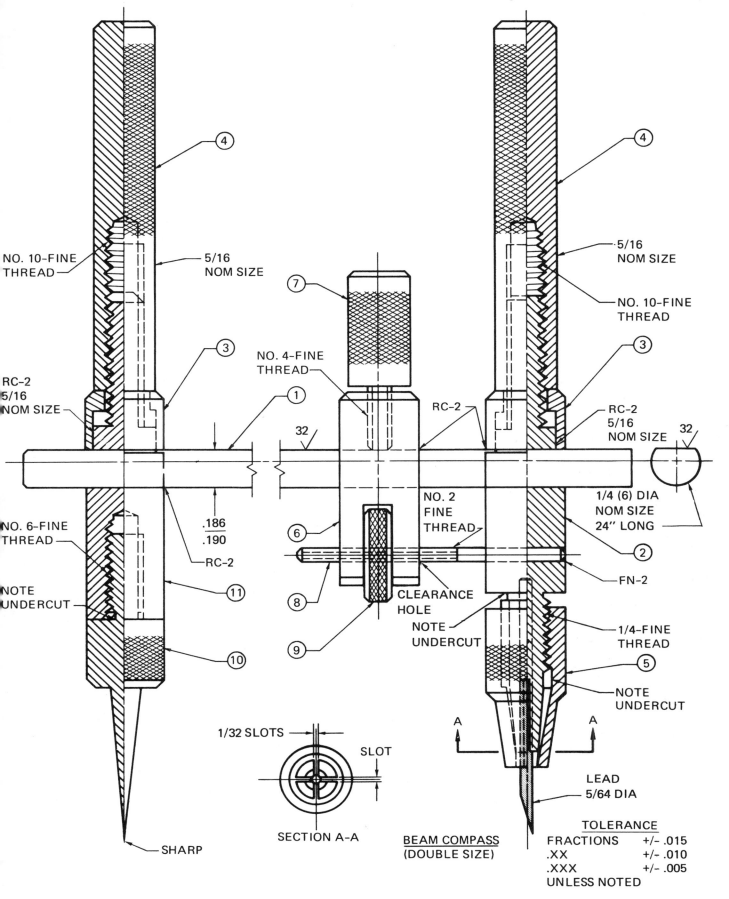

NO. 10-FINE THREAD

5/16 NOM SIZE

NO. 4-FINE THREAD

5/16 NOM SIZE

NO. 10-FINE THREAD

RC-2 5/16 NOM SIZE

RC-2

RC-2 5/16 NOM SIZE

32

32

.186 .190

RC-2

NO. 6-FINE THREAD

NO. 2 FINE THREAD

1/4 (6) DIA NOM SIZE 24" LONG

FN-2

NOTE UNDERCUT

CLEARANCE HOLE

NOTE UNDERCUT

1/4-FINE THREAD

NOTE UNDERCUT

SHARP

1/32 SLOTS

SLOT

A   A

LEAD 5/64 DIA

SECTION A-A

BEAM COMPASS (DOUBLE SIZE)

TOLERANCE	
FRACTIONS	+/- .015
.XX	+/- .010
.XXX	+/- .005
UNLESS NOTED	

Practice Exercise 8-2

Make an assembly drawing of the model airplane engine parts numbered 1 through 15. Parts 1 and 2 have been made into a subassembly in part 3. Compare your work to the answer on page 208. Save these drawings as they will be used to complete practice exercise 11-2 in Unit 11. The numbers inside the triangles, ⟨1⟩ , identify features to be used in Exercise 8-3. Note: These drawings are dimensioned in inches only.

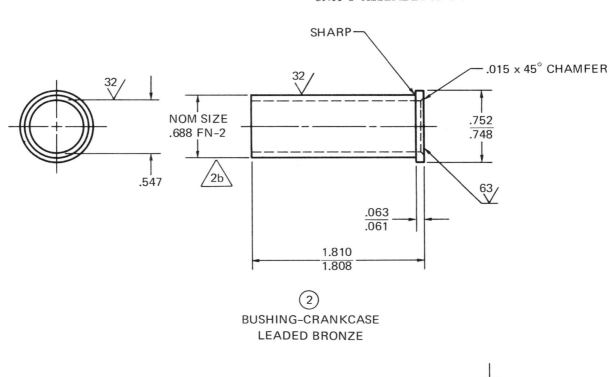

SHARP

.015 x 45° CHAMFER

32

32

NOM SIZE
.688 FN-2

.752
.748

2b

.547

63

.063
.061

1.810
1.808

② BUSHING–CRANKCASE
LEADED BRONZE

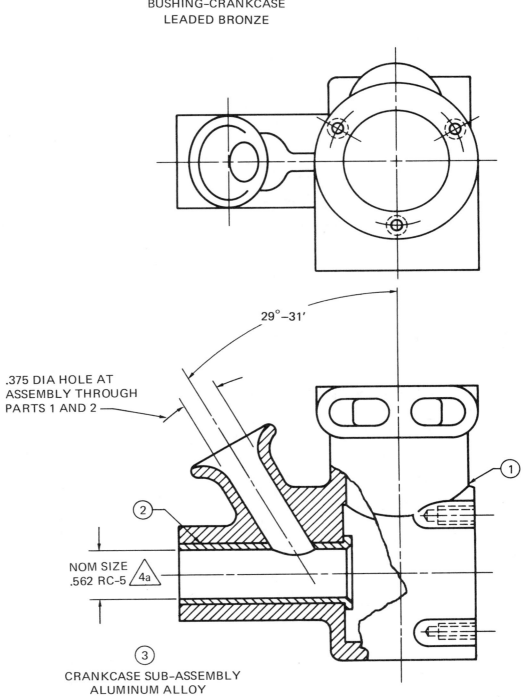

29°–31′

.375 DIA HOLE AT
ASSEMBLY THROUGH
PARTS 1 AND 2

①

②

NOM SIZE
.562 RC-5 ⚠4a

③ CRANKCASE SUB–ASSEMBLY
ALUMINUM ALLOY

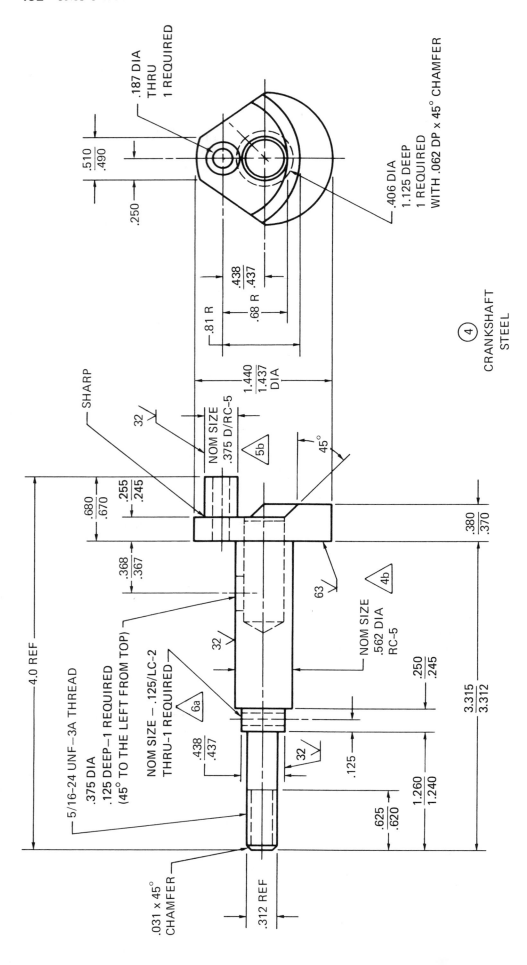

.187 DIA
THRU
1 REQUIRED

.510
.490

.250

.406 DIA
1.125 DEEP
1 REQUIRED
WITH .062 DP x 45° CHAMFER

.438
.437

.81 R

.68 R

1.440
1.437
DIA

SHARP

32

NOM SIZE
.375 D/RC-5

5b

45°

.255
.245

.680
.670

.380
.370

.368
.367

63

4b

NOM SIZE
.562 DIA
RC-5

32

.250
.245

4.0 REF

3.315
3.312

5/16-24 UNF-3A THREAD

.375 DIA
.125 DEEP-1 REQUIRED
(45° TO THE LEFT FROM TOP)

NOM SIZE - .125/LC-2
THRU-1 REQUIRED

6a

.438
.437

32

.125

1.260
1.240

.625
.620

.031 x 45°
CHAMFER

.312 REF

4

CRANKSHAFT
STEEL

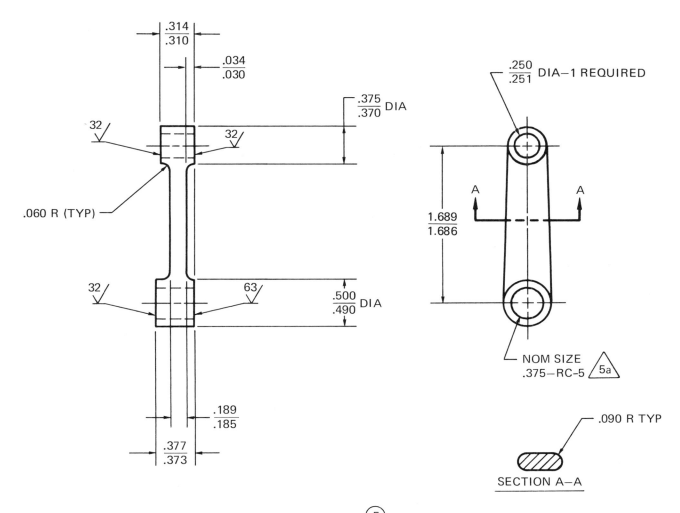

⑤
ROD-CONNECTING
ALUMINUM ALLOY

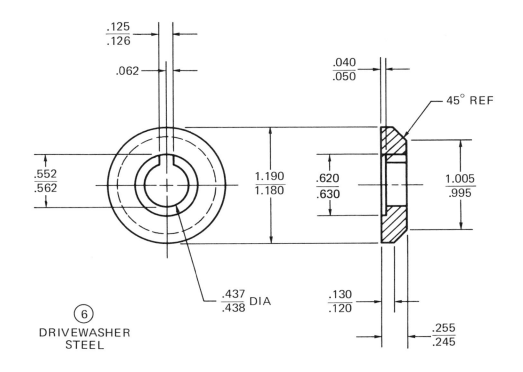

⑥
DRIVEWASHER
STEEL

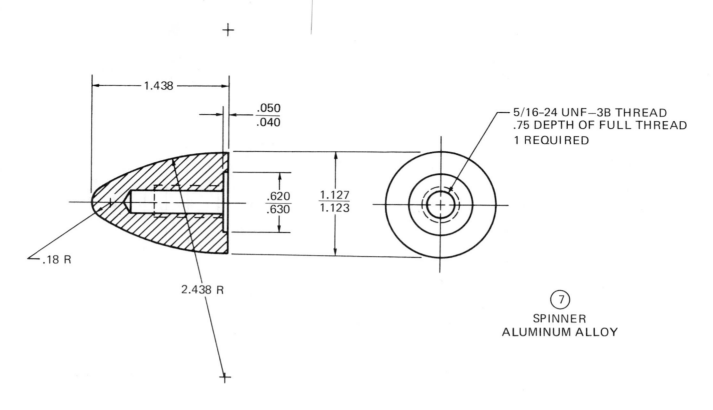

1.438

.050 / .040

5/16-24 UNF-3B THREAD
.75 DEPTH OF FULL THREAD
1 REQUIRED

.620 / .630

1.127 / 1.123

.18 R

2.438 R

⑦
SPINNER
ALUMINUM ALLOY

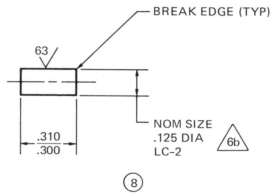

BREAK EDGE (TYP)

63

.310 / .300

NOM SIZE
.125 DIA
LC-2 △6b

⑧
PIN-DRIVEWASHER
(SCALE: DOUBLE SIZE)

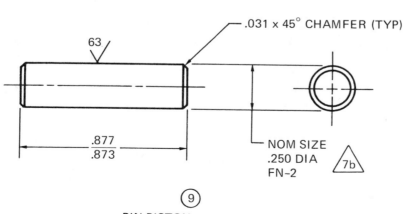

.031 x 45° CHAMFER (TYP)

63

.877 / .873

NOM SIZE
.250 DIA
FN-2 △7b

⑨
PIN PISTON
1010 STEEL
(SCALE: DOUBLE SIZE)

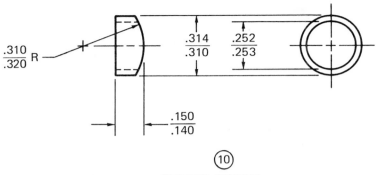

(10)
SPACER–PISTON
BRASS
(SCALE: DOUBLE SIZE)

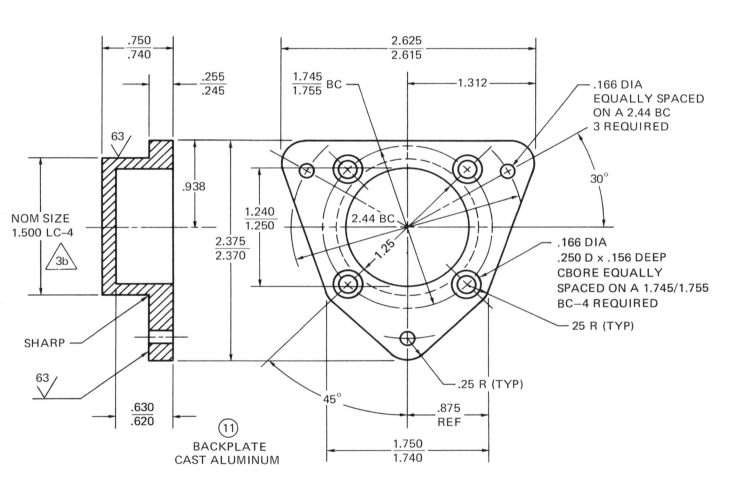

(11)
BACKPLATE
CAST ALUMINUM

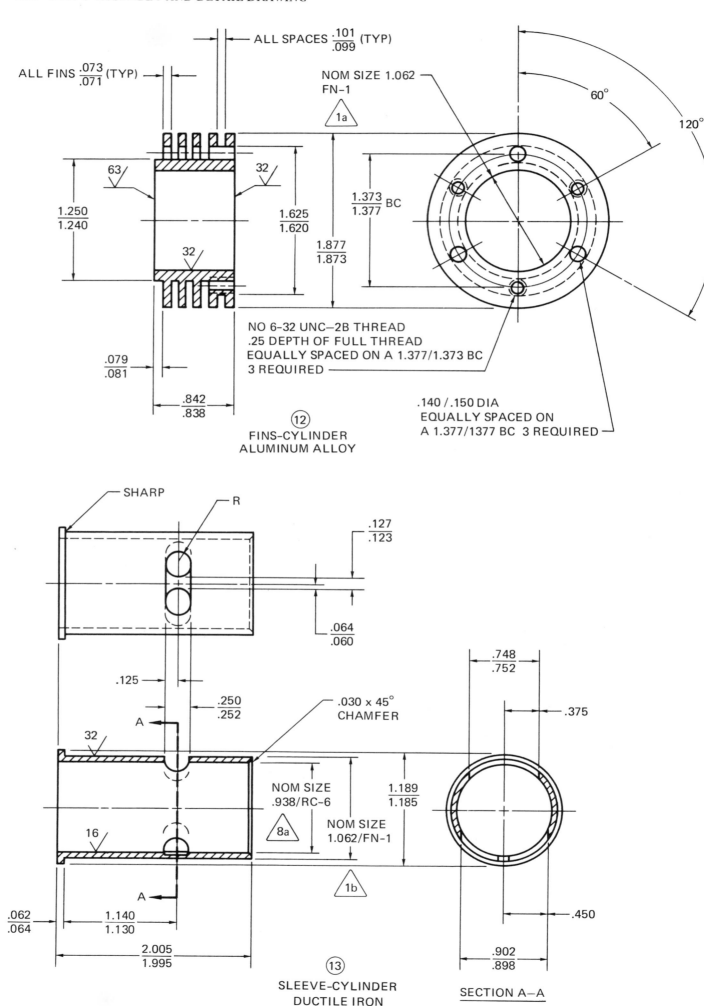

ALL SPACES $\frac{.101}{.099}$ (TYP)

ALL FINS $\frac{.073}{.071}$ (TYP)

NOM SIZE 1.062
FN-1

1a

60°

120°

63

32

32

$\frac{1.250}{1.240}$

$\frac{1.625}{1.620}$

$\frac{1.373}{1.377}$ BC

$\frac{1.877}{1.873}$

$\frac{.079}{.081}$

$\frac{.842}{.838}$

NO 6-32 UNC-2B THREAD
.25 DEPTH OF FULL THREAD
EQUALLY SPACED ON A 1.377/1.373 BC
3 REQUIRED

12

FINS-CYLINDER
ALUMINUM ALLOY

.140 / .150 DIA
EQUALLY SPACED ON
A 1.377/1377 BC 3 REQUIRED

SHARP

R

$\frac{.127}{.123}$

$\frac{.064}{.060}$

.125

$\frac{.250}{.252}$

.030 x 45°
CHAMFER

32

A

16

NOM SIZE
.938/RC-6

8a

NOM SIZE
1.062/FN-1

1b

$\frac{1.189}{1.185}$

$\frac{.748}{.752}$

.375

A

$\frac{.062}{.064}$

$\frac{1.140}{1.130}$

$\frac{2.005}{1.995}$

.450

$\frac{.902}{.898}$

13

SLEEVE-CYLINDER
DUCTILE IRON

SECTION A-A

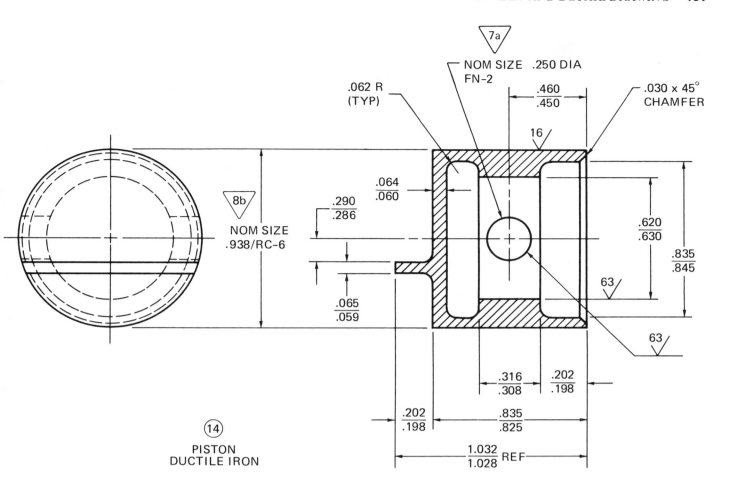

PISTON
DUCTILE IRON

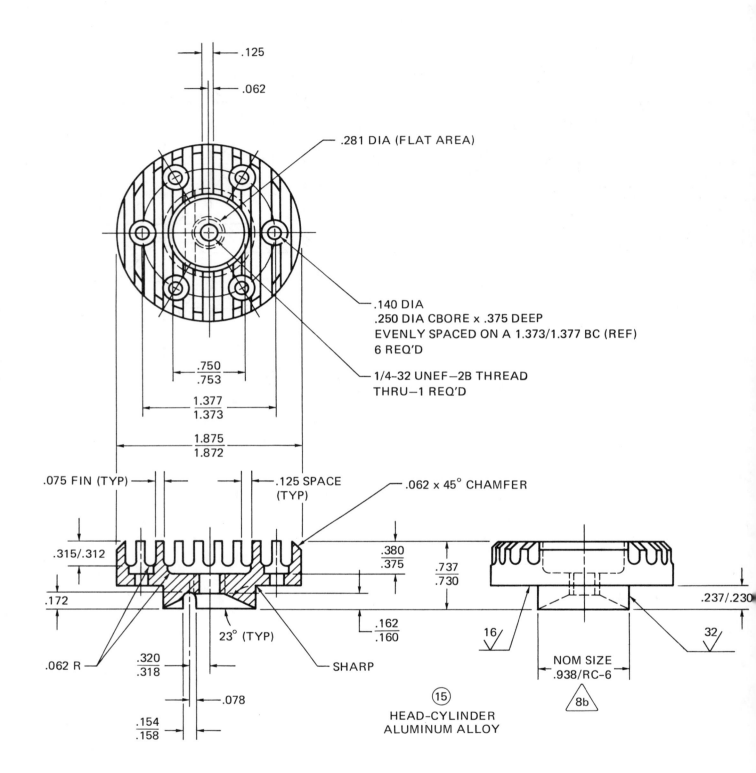

.125

.062

.281 DIA (FLAT AREA)

.140 DIA
.250 DIA CBORE x .375 DEEP
EVENLY SPACED ON A 1.373/1.377 BC (REF)
6 REQ'D

1/4–32 UNEF–2B THREAD
THRU–1 REQ'D

.750
.753

1.377
1.373

1.875
1.872

.075 FIN (TYP)

.125 SPACE
(TYP)

.062 x 45° CHAMFER

.315/.312

.380
.375

.737
.730

.172

.162
.160

.237/.230

23° (TYP)

16

32

.062 R

.320
.318

.078

SHARP

NOM SIZE
.938/RC-6

.154
.158

⑮

⚠8b

HEAD–CYLINDER
ALUMINUM ALLOY

Practice Exercise 8-3

Using the detail drawings for the model airplane engine, parts 1 through 15 in Exercise 8-2, and the required fit chart in Appendix C, calculate each fit for both the hole and shaft. Enter the answers in the spaces provided. Compare your work to the answers on page 209.

Prob.	Fit	Nom. Size	Answers Hole (a)	Shaft (b)	Part Nos.
△1	FN-1	1.062	——	——	1/12
△2	FN-2	.688	——	——	1/2
△3	LC-4	1.500	——	——	1/11
△4	RC-5	.562	——	——	3/4
△5	RC-5	.375	——	——	4/5
△6	LC-2	.125	——	——	4/8
△7	FN-2	.250	——	——	9/14
△8	RC-6	.938	——	——	13/14/15

 Layout Problem. What position will the piston (part 15) be in when the .375-inch diameter hole in the crankshaft (part 4) is in line with the .375-inch diameter hole in the crankshaft (part 3)? Note that the .375-inch diameter hole is located 45 degrees to the left (from top dead center) per detail drawing. Draw the layout.

Required Fasteners. What size fillister-head machine screws are used to hold all parts together? There are three different sizes, ten screws total. List the three sizes using the standard callout for fasteners.

Practice Exercise 8-4

Using the full-size design layout, detail each part. Use correct tolerancing on all fits. Use correct callouts on all notes and threads. Follow all dimensioning standards. Compare your work to the answers on pages 210 through 214. Note that the ball bearing data given (this information is usually found in a manufacturer's handbook) uses the same tolerance as outlined in practice exercise 8-1. Also note that this drawing is dimensioned in inches and millimetres.

Problem: To apply rotation inside an airtight sealed chamber from the outside of the chamber. Note that the flange (#1) bolts to a matching flange on the chamber.

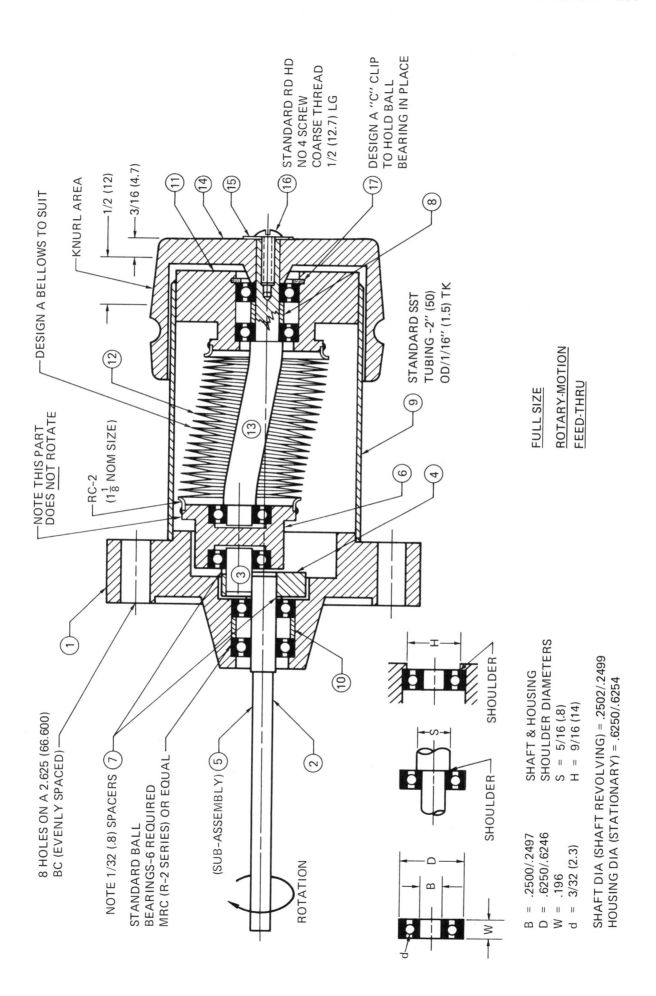

STANDARD RD HD
NO 4 SCREW
COARSE THREAD
1/2 (12.7) LG

DESIGN A "C" CLIP
TO HOLD BALL
BEARING IN PLACE

1/2 (12)

3/16 (4.7)

KNURL AREA

DESIGN A BELLOWS TO SUIT

NOTE THIS PART
DOES NOT ROTATE

RC-2
(1 1/8 NOM SIZE)

STANDARD SST
TUBING -2'' (50)
OD/1/16'' (1.5) TK

8 HOLES ON A 2.625 (66.600)
BC (EVENLY SPACED)

NOTE 1/32 (.8) SPACERS

STANDARD BALL
BEARINGS-6 REQUIRED
MRC (R-2 SERIES) OR EQUAL

(SUB-ASSEMBLY)

ROTATION

FULL SIZE
ROTARY-MOTION
FEED-THRU

SHOULDER

SHOULDER

SHOULDER

SHAFT & HOUSING
SHOULDER DIAMETERS
S = 5/16 (.8)
H = 9/16 (14)

B = .2500/.2497
D = .6250/.6246
W = .196
d = 3/32 (2.3)

SHAFT DIA (SHAFT REVOLVING) = .2502/.2499
HOUSING DIA (STATIONARY) = .6250/.6254

Practice Exercise 8-5

Very carefully, using the full-size design layout, detail each part circled. Use standard drafting methods and the tolerances outlined in practice exercise 8-1. Compare your work to the answers on pages 215 through 224. Note that the drawings are dimensioned in inches and millimetres.

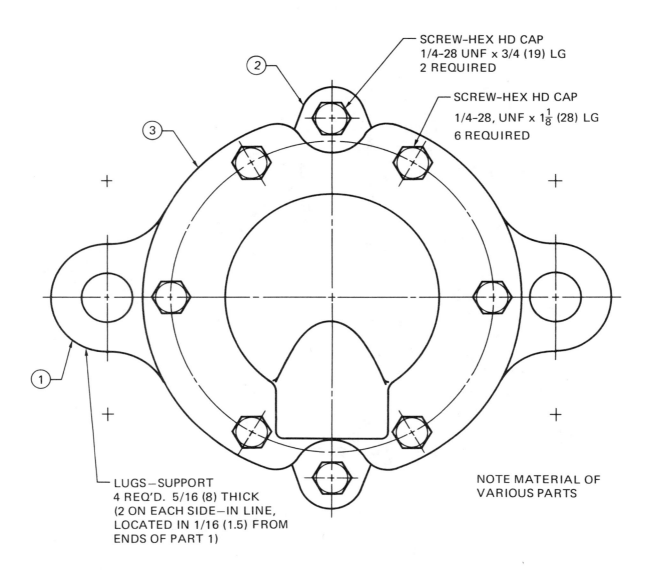

SCREW–HEX HD CAP
1/4-28 UNF x 3/4 (19) LG
2 REQUIRED

SCREW–HEX HD CAP
1/4-28, UNF x $1\frac{1}{8}$ (28) LG
6 REQUIRED

LUGS–SUPPORT
4 REQ'D. 5/16 (8) THICK
(2 ON EACH SIDE–IN LINE,
LOCATED IN 1/16 (1.5) FROM
ENDS OF PART 1)

NOTE MATERIAL OF
VARIOUS PARTS

ROTARY PRESSURE JOINT

From a basic design of THE JOHNSON CORP.
Michigan–49093

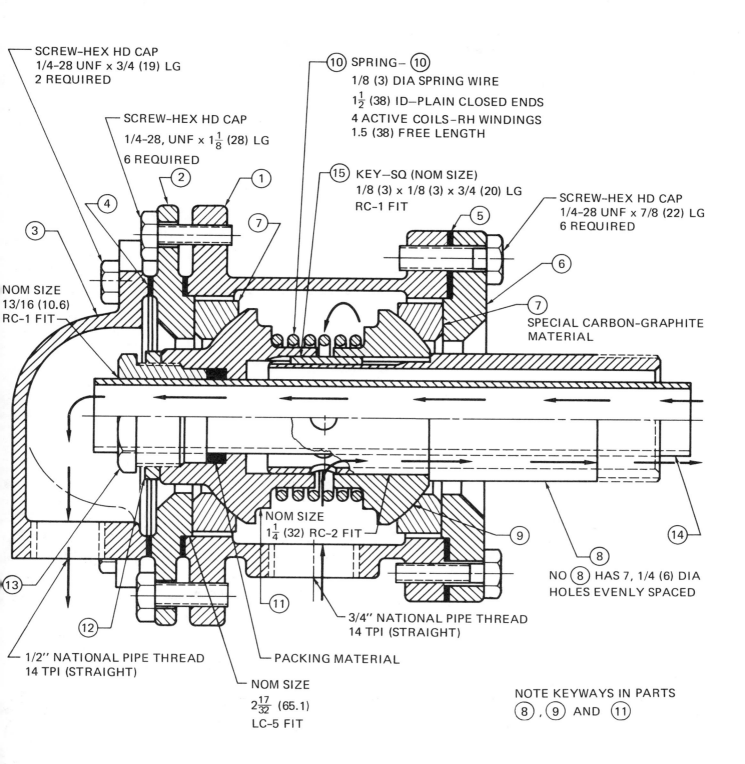

SCREW–HEX HD CAP
1/4–28 UNF x 3/4 (19) LG
2 REQUIRED

SCREW–HEX HD CAP
1/4–28, UNF x 1 1/8 (28) LG
6 REQUIRED

NOM SIZE
13/16 (10.6)
RC–1 FIT

⑩ SPRING– ⑩
1/8 (3) DIA SPRING WIRE
1 1/2 (38) ID–PLAIN CLOSED ENDS
4 ACTIVE COILS–RH WINDINGS
1.5 (38) FREE LENGTH

⑮ KEY–SQ (NOM SIZE)
1/8 (3) x 1/8 (3) x 3/4 (20) LG
RC–1 FIT

SCREW–HEX HD CAP
1/4–28 UNF x 7/8 (22) LG
6 REQUIRED

SPECIAL CARBON–GRAPHITE
MATERIAL

NOM SIZE
1 1/4 (32) RC–2 FIT

3/4″ NATIONAL PIPE THREAD
14 TPI (STRAIGHT)

NO ⑧ HAS 7, 1/4 (6) DIA
HOLES EVENLY SPACED

1/2″ NATIONAL PIPE THREAD
14 TPI (STRAIGHT)

PACKING MATERIAL

NOM SIZE
2 17/32 (65.1)
LC–5 FIT

NOTE KEYWAYS IN PARTS
⑧ , ⑨ AND ⑪

ANSWERS TO PRACTICE EXERCISES

Exercise 8-1. A design layout is drawn to approximate size. It is up to the drafter to redraw each part using standard size material and standard tools wherever possible. The drafter must calculate all allowances between mating parts and add any finish marks where needed.

Part 1

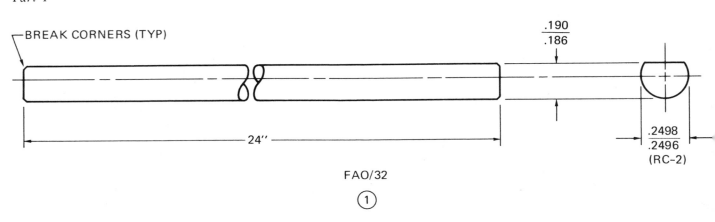

FAO/32

①

Part 2

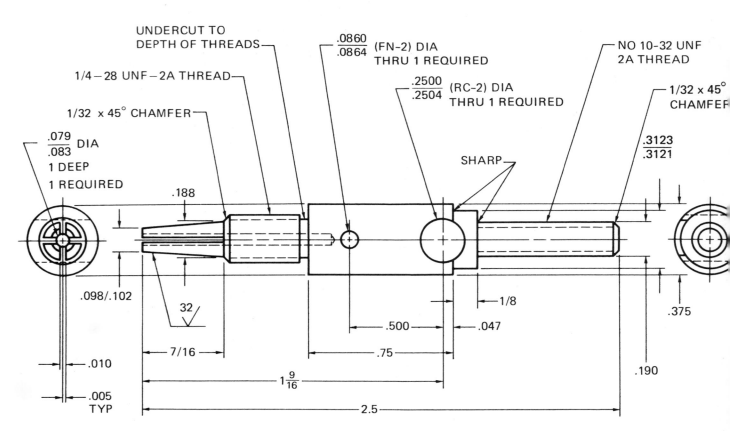

②

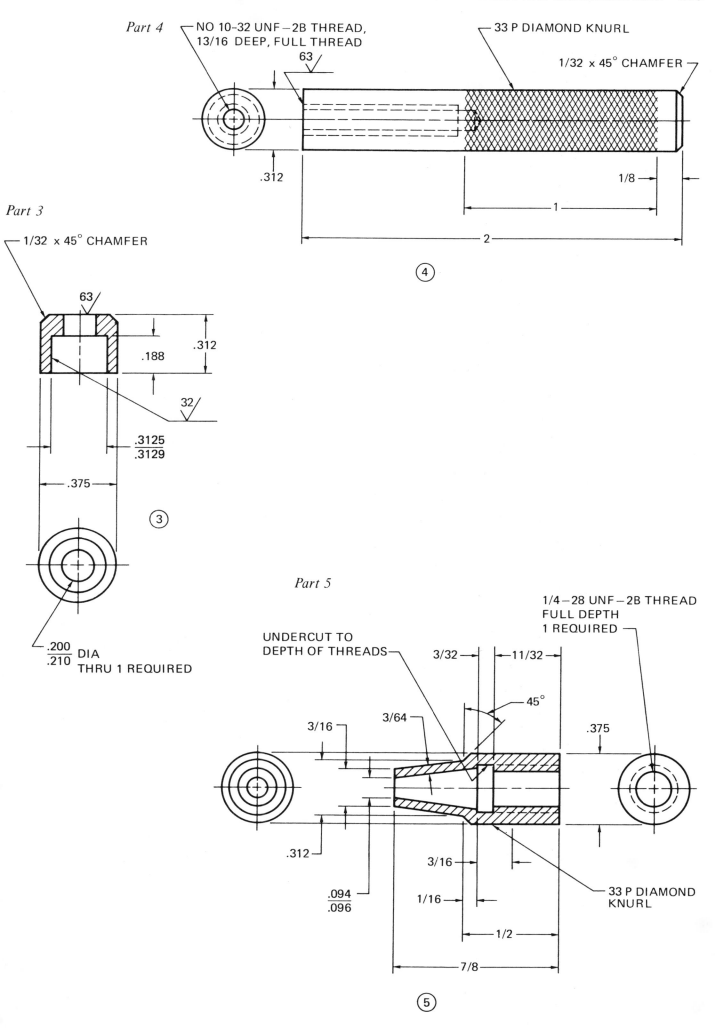

Part 4 — NO 10–32 UNF –2B THREAD, 13/16 DEEP, FULL THREAD

33 P DIAMOND KNURL

1/32 × 45° CHAMFER

63

.312

1/8

1

2

④

Part 3

1/32 × 45° CHAMFER

63

.312

.188

32

.3125 / .3129

.375

③

.200 / .210 DIA THRU 1 REQUIRED

Part 5

1/4 –28 UNF –2B THREAD FULL DEPTH 1 REQUIRED

UNDERCUT TO DEPTH OF THREADS

3/32

11/32

3/64

45°

3/16

.375

.312

3/16

.094 / .096

1/16

1/2

7/8

33 P DIAMOND KNURL

⑤

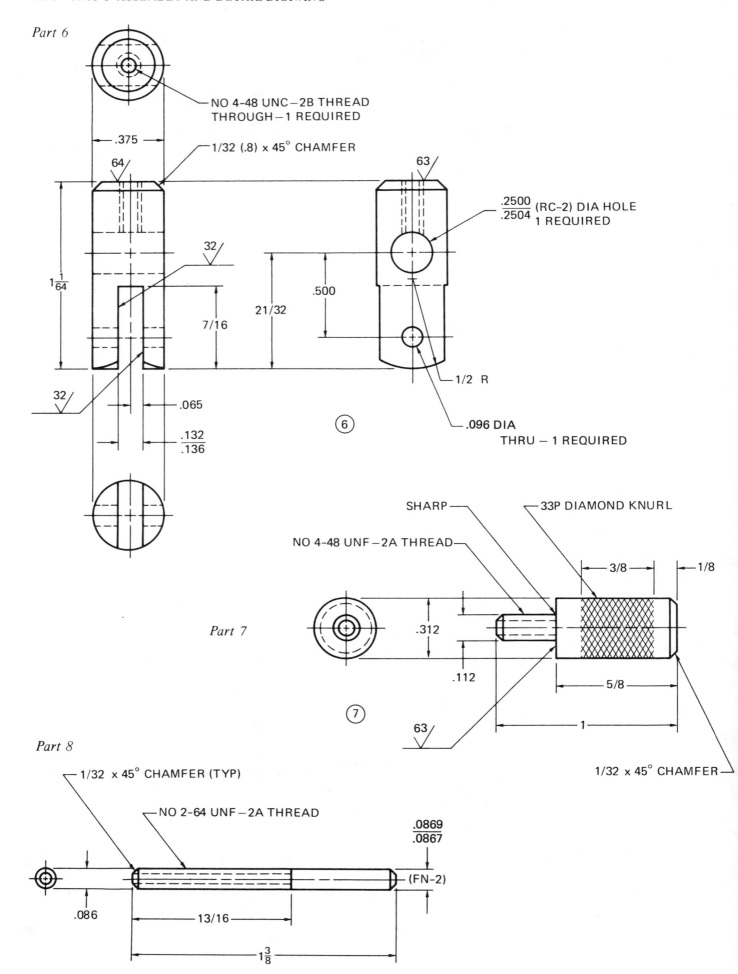

Part 6

NO 4-48 UNC–2B THREAD
THROUGH–1 REQUIRED

1/32 (.8) x 45° CHAMFER

64

.375

32

$1\frac{1}{64}$

7/16

21/32

32

.065

.132
.136

6

63

.2500/.2504 (RC–2) DIA HOLE
1 REQUIRED

.500

1/2 R

.096 DIA
THRU – 1 REQUIRED

SHARP

33P DIAMOND KNURL

NO 4-48 UNF – 2A THREAD

3/8

1/8

Part 7

.312

.112

5/8

1

63

1/32 x 45° CHAMFER

7

Part 8

1/32 x 45° CHAMFER (TYP)

NO 2-64 UNF – 2A THREAD

.0869
.0867

.086

(FN–2)

13/16

$1\frac{3}{8}$

8

Part 9

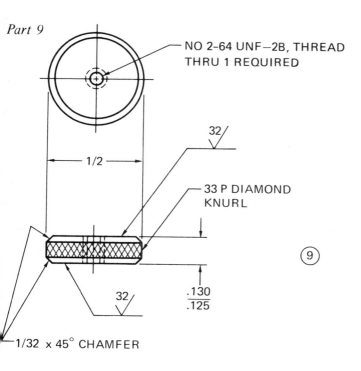

NO 2–64 UNF–2B, THREAD
THRU 1 REQUIRED

32/

33 P DIAMOND
KNURL

1/2

9

.130
.125

32/

1/32 x 45° CHAMFER

Part 10

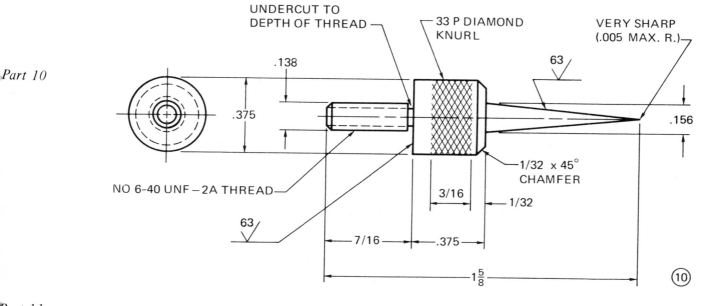

UNDERCUT TO
DEPTH OF THREAD

33 P DIAMOND
KNURL

VERY SHARP
(.005 MAX. R.)

63/

.138

.375

.156

NO 6-40 UNF – 2A THREAD

1/32 x 45°
CHAMFER

63/

3/16

1/32

7/16

.375

1⅝

10

Part 11

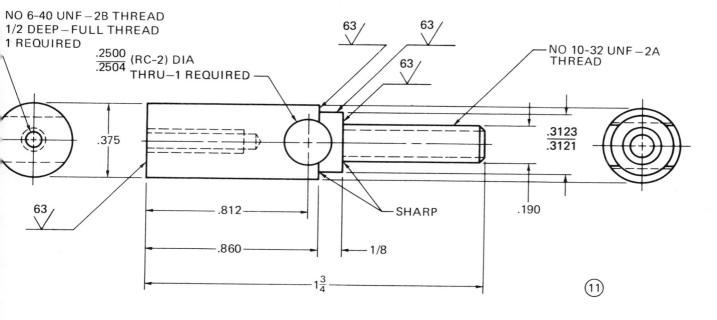

NO 6-40 UNF – 2B THREAD
1/2 DEEP – FULL THREAD
1 REQUIRED

.2500
.2504 (RC-2) DIA
THRU–1 REQUIRED

63/   63/

63/

NO 10-32 UNF – 2A
THREAD

.375

.3123
.3121

63/

SHARP

.190

.812

.860

1/8

1¾

11

Exercise 8-2

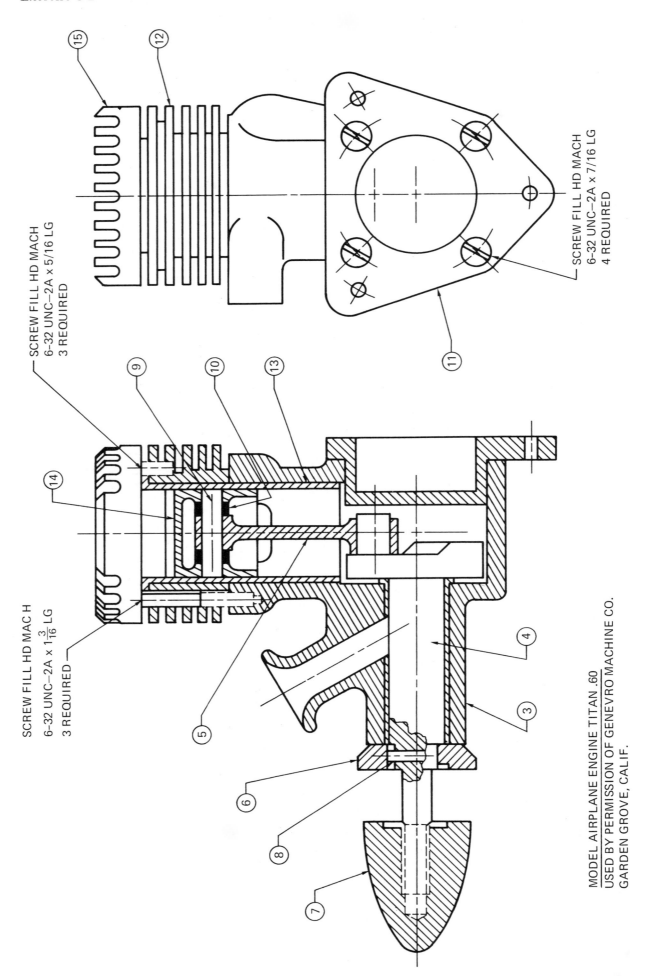

SCREW FILL HD MACH
6-32 UNC–2A x 5/16 LG
3 REQUIRED

SCREW FILL HD MACH
6-32 UNC–2A x 7/16 LG
4 REQUIRED

SCREW FILL HD MAC H
6-32 UNC–2A x 1$\frac{3}{16}$ LG
3 REQUIRED

MODEL AIRPLANE ENGINE TITAN .60
USED BY PERMISSION OF GENEVRO MACHINE CO.
GARDEN GROVE, CALIF.

Practice Exercise 8-3

NOTE: The correct way to call out the hole limits is with the small limit figure on top. The correct way to call out the shaft limit is with the large limit figure on top, as noted in the answer column below.

| Prob. | Fit | Nom. Size | Answers | | Part Nos. |
			Hole (a)	Shaft (b)	
1	FN-1	1.062	1.0620 1.0625	1.0632 1.0628	1/12
2	FN-2	.688	.6880 .6887	.6896 .6892	1/2
3	LC-4	1.500	1.5000 1.504	1.5000 1.4975	1/11
4	RC-5	.562	.5620 .5630	.5608 .5601	3/4
5	RC-5	.375	.3750 .3759	.3740 .3734	4/5
6	LC-2	.125	.1250 .1255	.1250 .1247	4/8
7	FN-2	.250	.2500 .2506	.2514 .2510	9/14
8	RC-6	.938	.9380 .9400	.9364 .9352	13/14/15

9   **Layout Problem.** The problem could be done by math calculations or by a simple layout as illustrated:

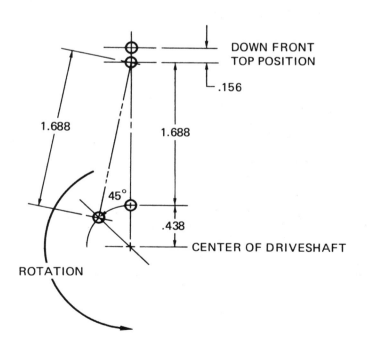

10   **Required Fasteners.** Use the correct callout noted:

    A.   Screw fill. hd. mach. 6-32 UNC-2A × 1 3/16 LG.-3 REQ'D. (fasten head)

    B.   Screw-fill. hd. mach. 6-32 UNC-2A × 5/16 LG.-3 REQ'D. (fasten head)

    C.   Screw fill. hd. mach. 6-32 UNC-2A × 7/16 LG.-4 REQ'D (fasten backplate).

Practice Exercise 8-4

Part 1

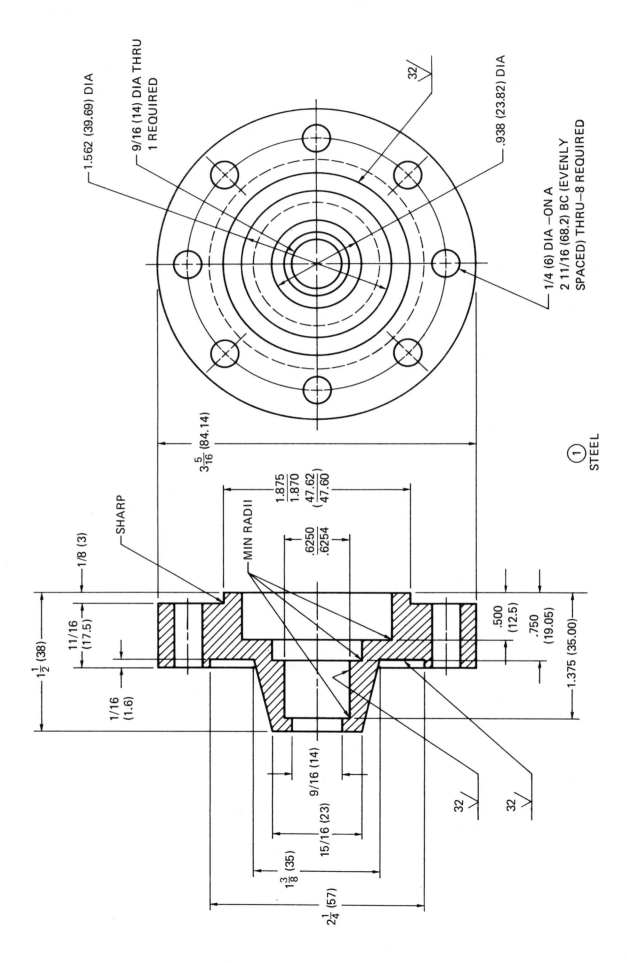

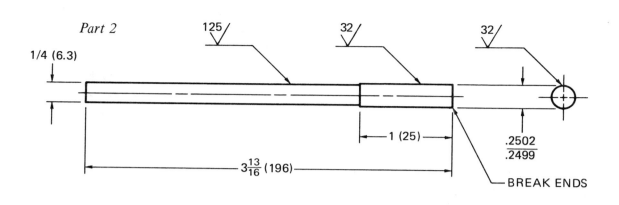

Part 2

125 /   32 /   32 /

1/4 (6.3)

1 (25)

.2502
.2499

$3\frac{13}{16}$ (196)

BREAK ENDS

②

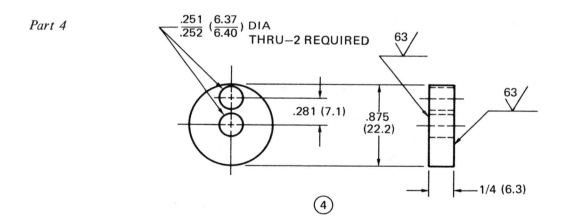

Part 3

32 /   BREAK ENDS   32 /

7/16
(11)

.2502
.2499

③

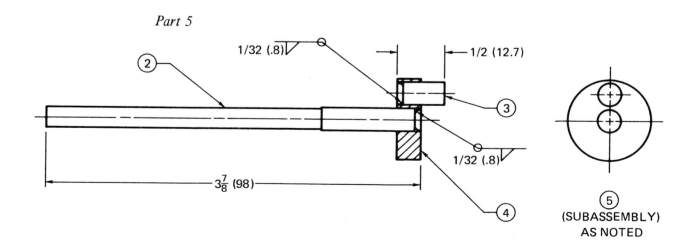

Part 4

$\frac{.251}{.252}$ ($\frac{6.37}{6.40}$) DIA
THRU−2 REQUIRED

63 /

63 /

.281 (7.1)

.875
(22.2)

1/4 (6.3)

④

Part 5

②

1/32 (.8) /   1/2 (12.7)

③

$3\frac{7}{8}$ (98)

1/32 (.8) /

④

⑤
(SUBASSEMBLY)
AS NOTED

Part 6

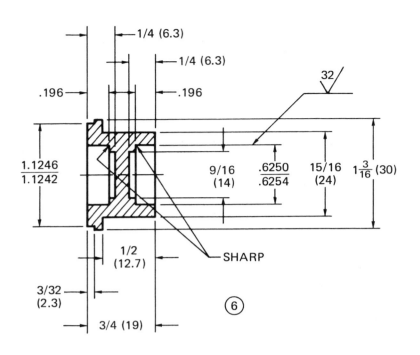

Part 7

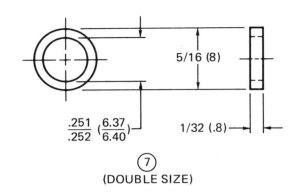

⑦
(DOUBLE SIZE)

Part 8

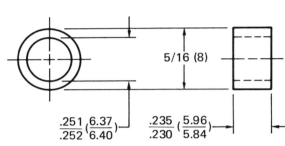

⑧
(DOUBLE SIZE)
STEEL

Part 9

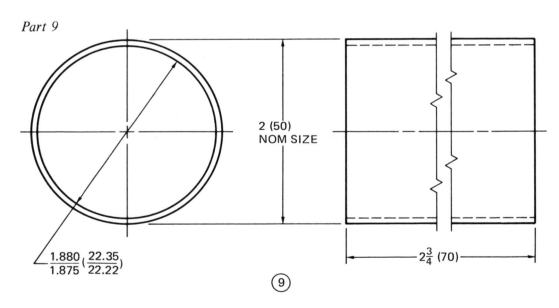

⑨

Part 10

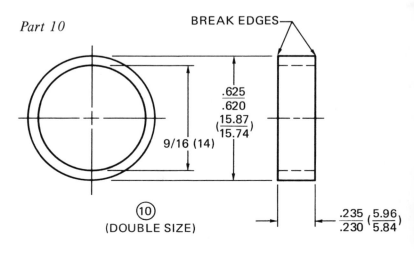

⑩
(DOUBLE SIZE)

Part 11

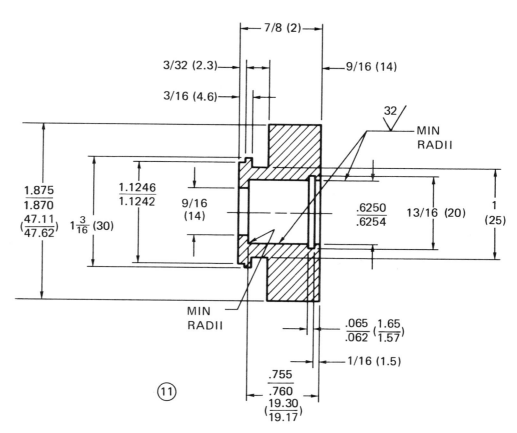

⑪

Part 12

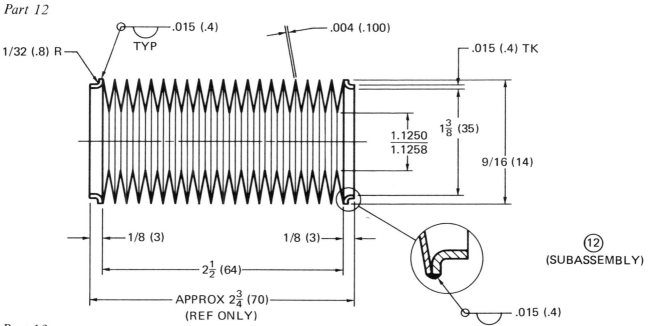

⑫
(SUBASSEMBLY)

Part 13

⑬

Part 14

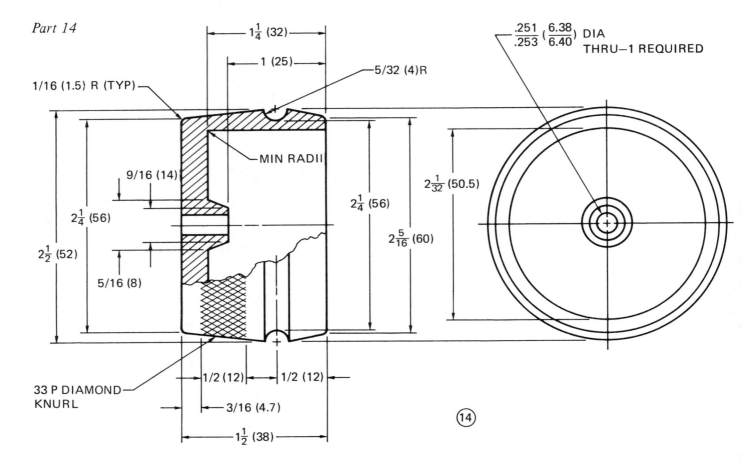

$2\frac{1}{32}$ (50.5)

.251/.253 $\left(\frac{6.38}{6.40}\right)$ DIA
THRU–1 REQUIRED

$1\frac{1}{4}$ (32)

1 (25)

5/32 (4)R

1/16 (1.5) R (TYP)

MIN RADII

9/16 (14)

$2\frac{1}{4}$ (56)

$2\frac{1}{2}$ (52)

5/16 (8)

$2\frac{1}{4}$ (56)

$2\frac{5}{16}$ (60)

33 P DIAMOND KNURL

1/2 (12) 1/2 (12)

3/16 (4.7)

$1\frac{1}{2}$ (38)

(14)

Part 15

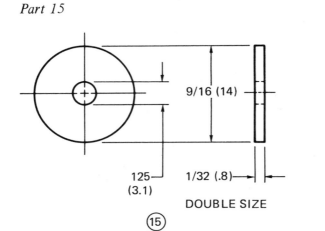

9/16 (14)

125 (3.1)

1/32 (.8)

DOUBLE SIZE

(15)

Part 17

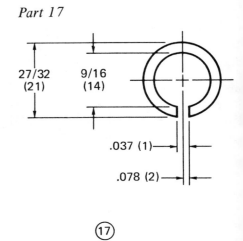

27/32 (21)

9/16 (14)

.037 (1)

.078 (2)

(17)

Practice Exercise 8-5

Part 1

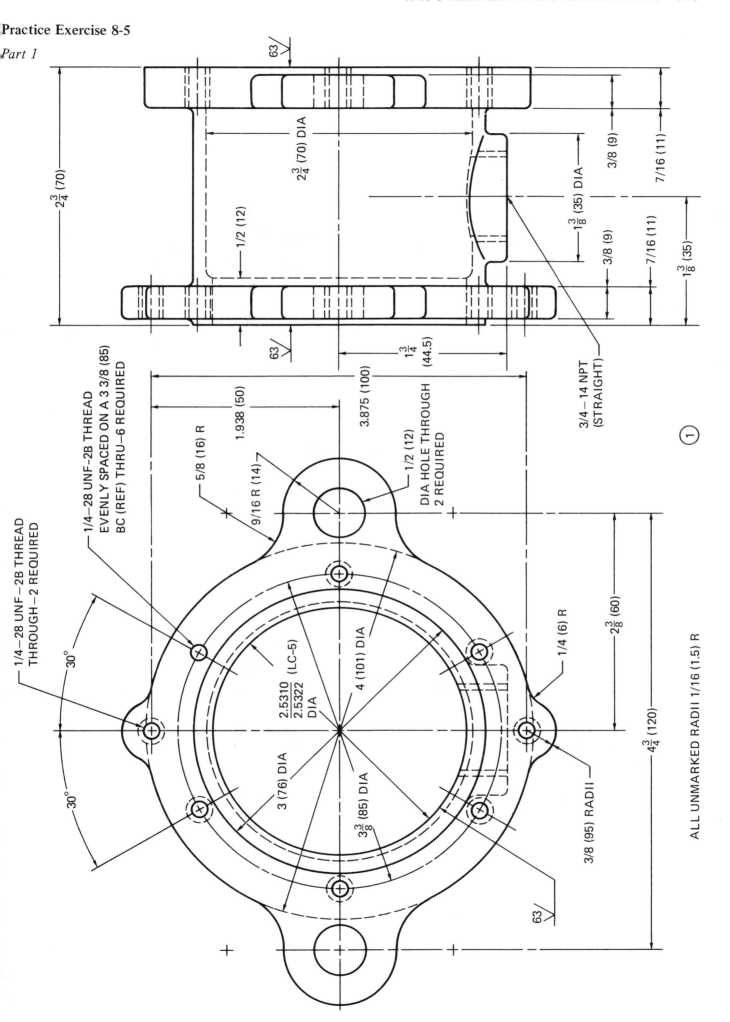

Part 2

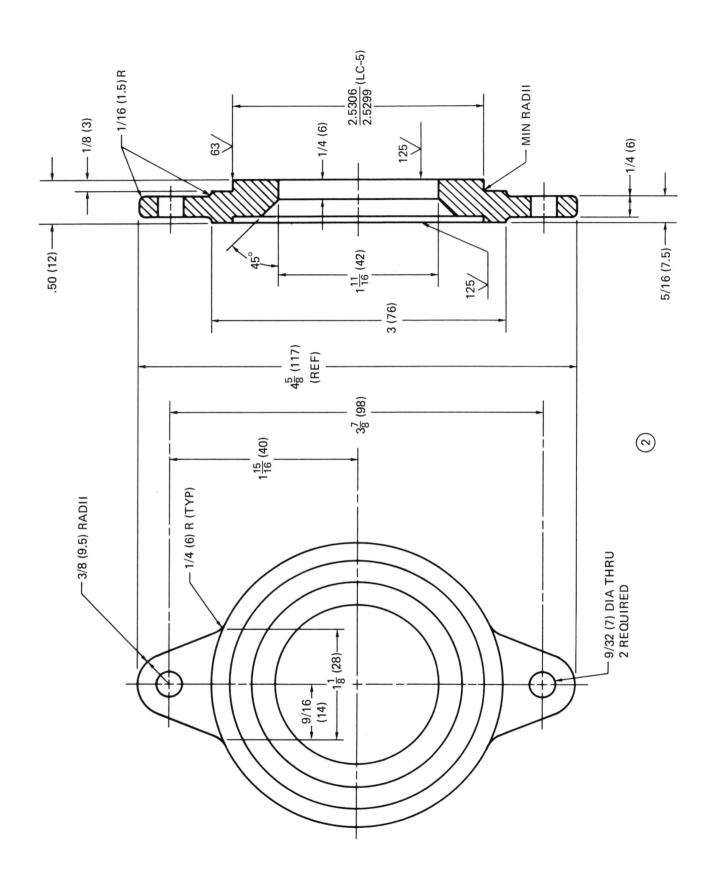

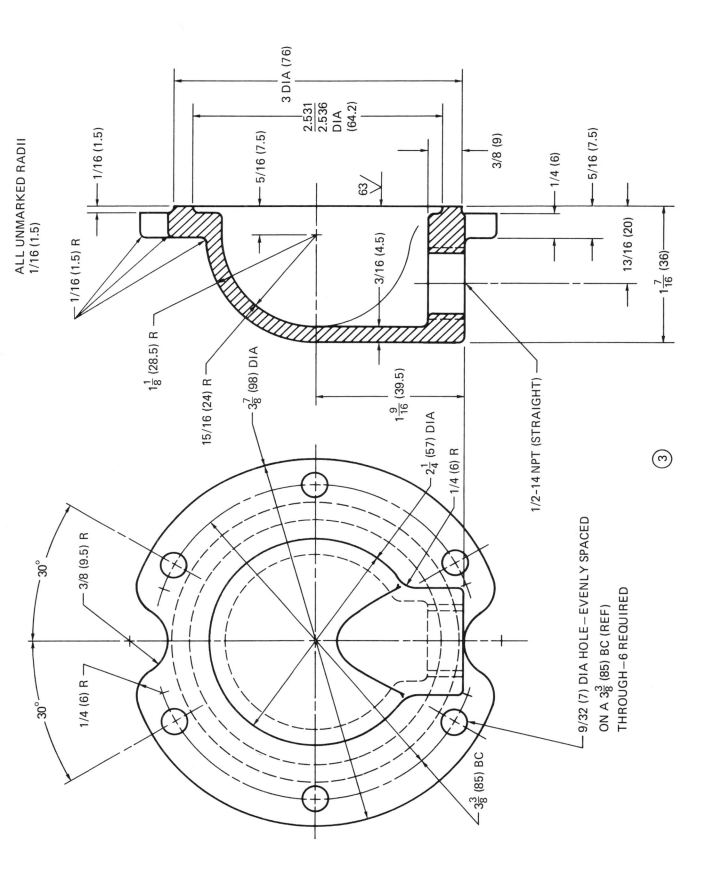

ALL UNMARKED RADII
1/16 (1.5)

3 DIA (76)

$\dfrac{2.531}{2.536}$ DIA (64.2)

1/16 (1.5)

1/16 (1.5)

5/16 (7.5)

3/8 (9)

1/4 (6)

5/16 (7.5)

63

1/16 (1.5) R

3/16 (4.5)

13/16 (20)

$1\frac{7}{16}$ (36)

$1\frac{1}{8}$ (28.5) R

15/16 (24) R

$3\frac{7}{8}$ (98) DIA

$1\frac{9}{16}$ (39.5)

$2\frac{1}{4}$ (57) DIA

1/4 (6) R

1/2-14 NPT (STRAIGHT)

③

30°

3/8 (9.5) R

30°

1/4 (6) R

9/32 (7) DIA HOLE–EVENLY SPACED
ON A $3\frac{3}{8}$ (85) BC (REF)
THROUGH–6 REQUIRED

$3\frac{3}{8}$ (85) BC

Part 4

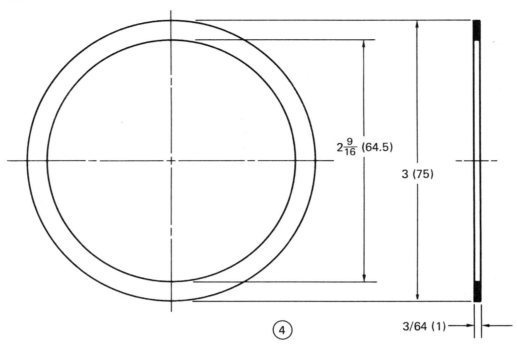

$2\frac{9}{16}$ (64.5)

3 (75)

④

3/64 (1)

Part 5

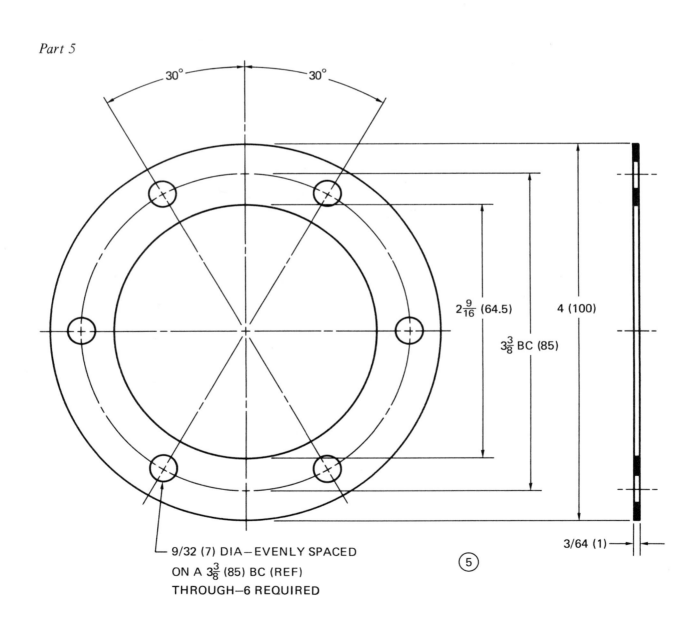

30° 30°

$2\frac{9}{16}$ (64.5)

4 (100)

$3\frac{3}{8}$ BC (85)

9/32 (7) DIA—EVENLY SPACED
ON A $3\frac{3}{8}$ (85) BC (REF)
THROUGH—6 REQUIRED

3/64 (1)

⑤

Part 6

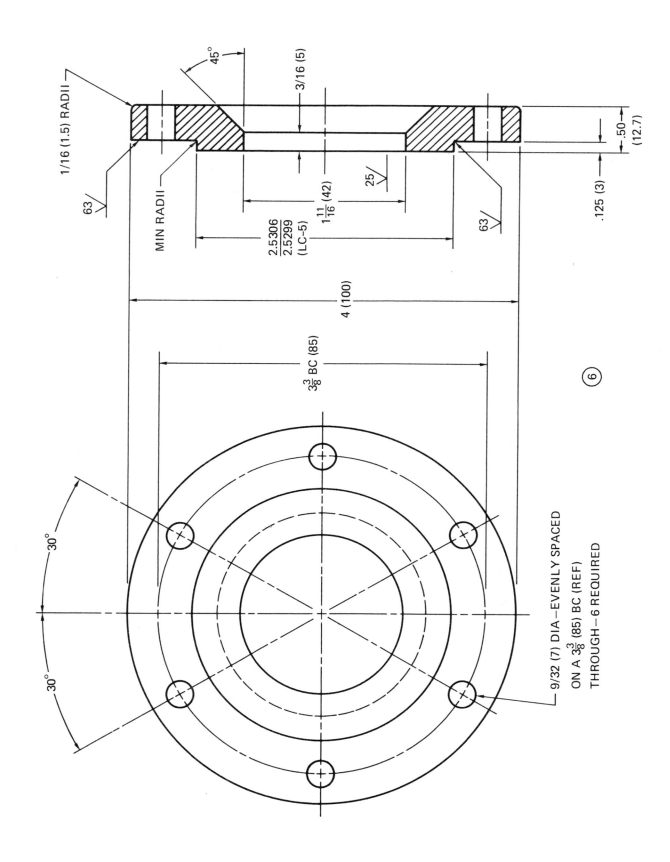

45°

3/16 (5)

1/16 (1.5) RADII

63

MIN RADII

.50
(12.7)

.125 (3)

25

63

$1\frac{11}{16}$ (42)

$\frac{2.5306}{2.5299}$ (LC-5)

4 (100)

$3\frac{3}{8}$ BC (85)

⑥

30°

30°

9/32 (7) DIA–EVENLY SPACED
ON A $3\frac{3}{8}$ (85) BC (REF)
THROUGH–6 REQUIRED

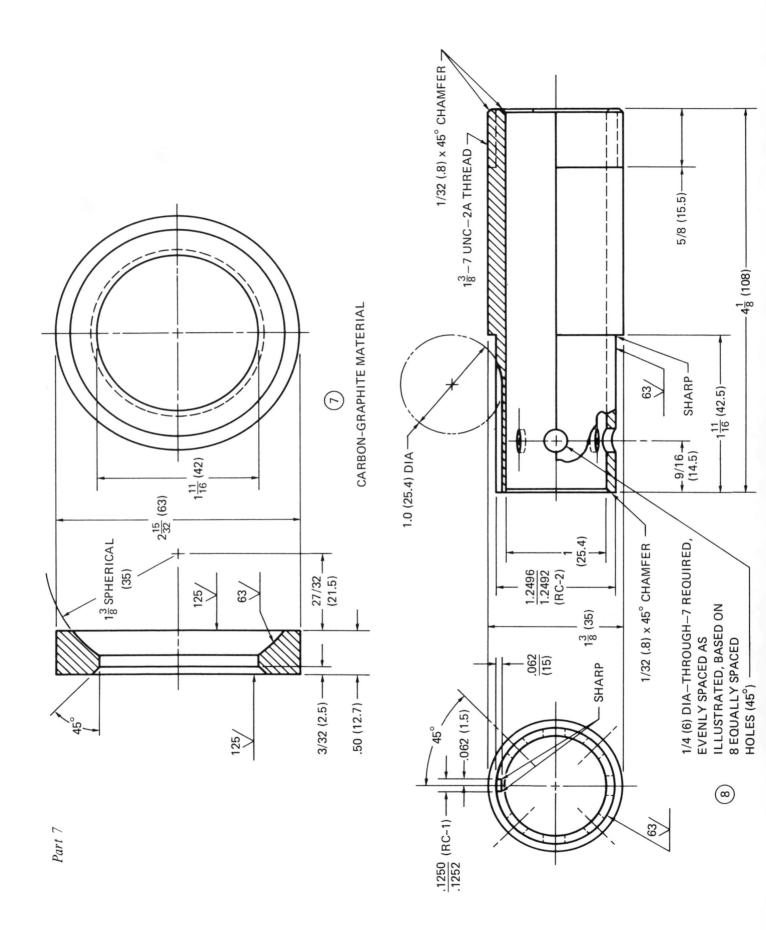

Part 7

Part 8

Part 9

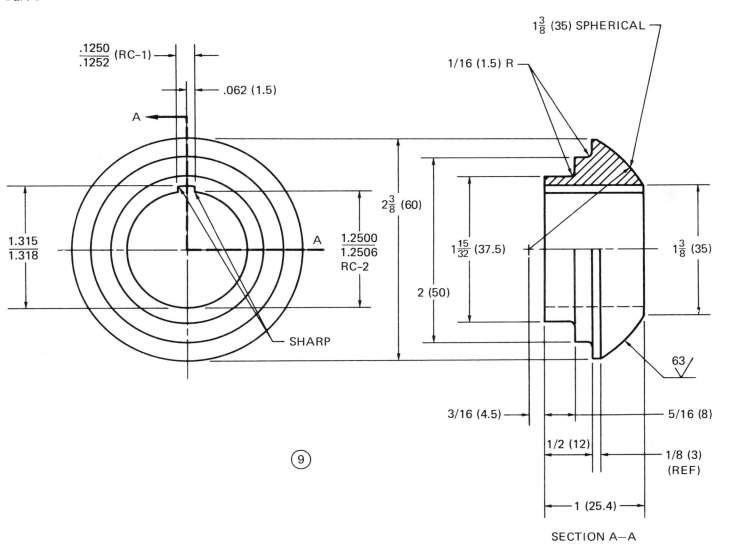

SECTION A–A

Part 10

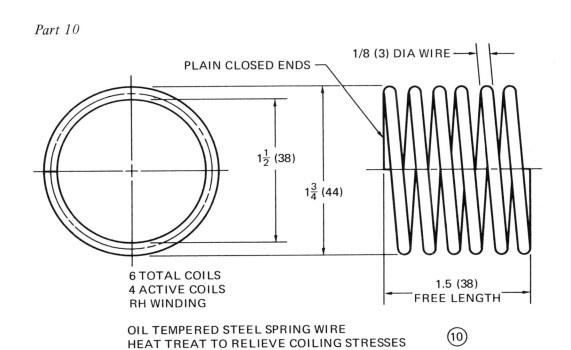

Part 11

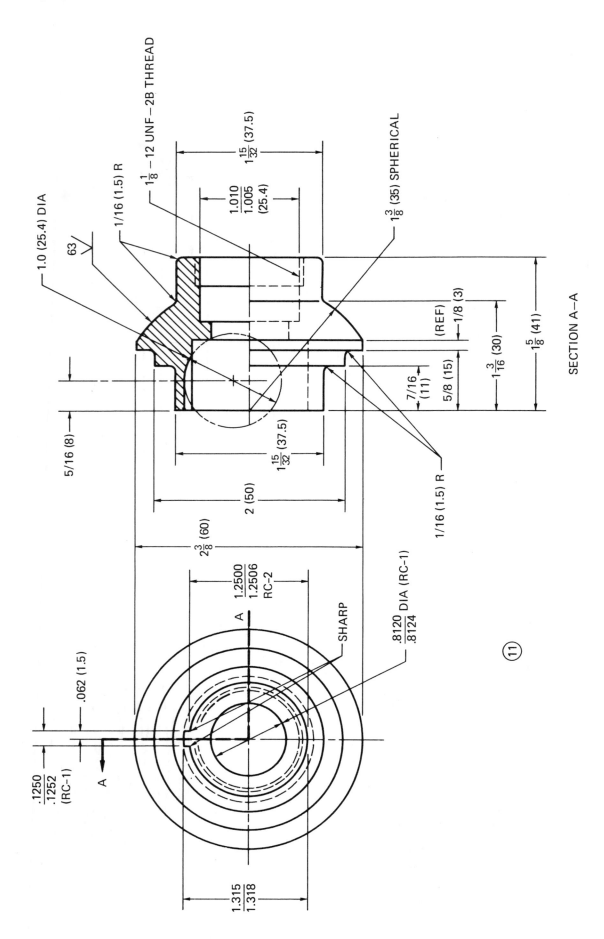

SECTION A–A

1⅛ – 12 UNF – 2B THREAD

1/16 (1.5) R

1.0 (25.4) DIA

63

1⁵⁄₃₂ (37.5)

1.010 / 1.005 (25.4)

1⅜ (35) SPHERICAL

(REF) 1/8 (3)

1⁵⁄₈ (41)

1³⁄₁₆ (30)

7/16 (11)

5/8 (15)

1/16 (1.5) R

5/16 (8)

1⁵⁄₃₂ (37.5)

2 (50)

2³⁄₈ (60)

.062 (1.5)

.1250 / .1252 (RC-1)

A

A

1.2500 / 1.2506 RC-2

SHARP

.8120 / .8124 DIA (RC-1)

1.315 / 1.318

⑪

Part 12

Part 13

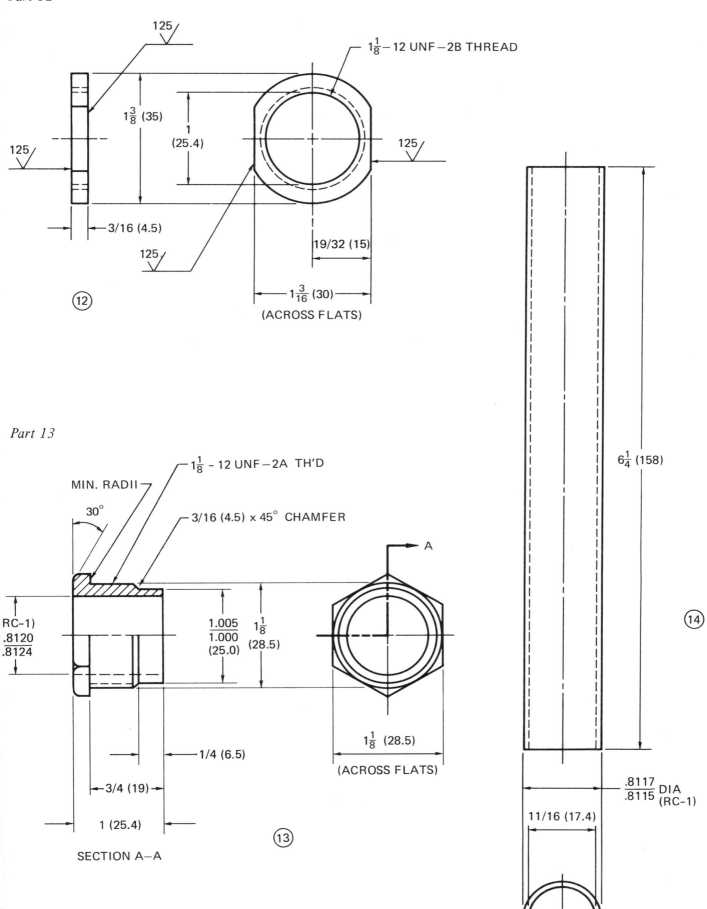

Part 14

Part 15

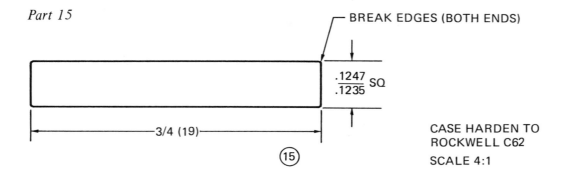

BREAK EDGES (BOTH ENDS)

$\frac{.1247}{.1235}$ SQ

3/4 (19)

(15)

CASE HARDEN TO
ROCKWELL C62
SCALE 4:1

Before proceeding to the next unit:

_____ Instructor's approval

_____ Progress plotted

UNIT 9

ELECTRONIC COMPONENTS

OBJECTIVE

The student will learn what various electronic components do and to identify schematic symbols.

PRETEST

45-minute time limit

Questions

1. Explain briefly how an inductor coil works.

2. If there are 1200 turns in the primary winding of a transformer with 110 volts coming into it, and 600 turns on the secondary winding, what is the output?

3. What is an integrated circuit?

4. What is the difference between electricity and electronics?

5. What are the three kinds of circuits?

6. How does a relay switch work? What is the advantage of using one?

7. How does a triode tube work? What does it do?

8. What does a diode do? What kind of tube can do the same thing?

Identification

Identify what each symbol represents and place its name in the space provided.

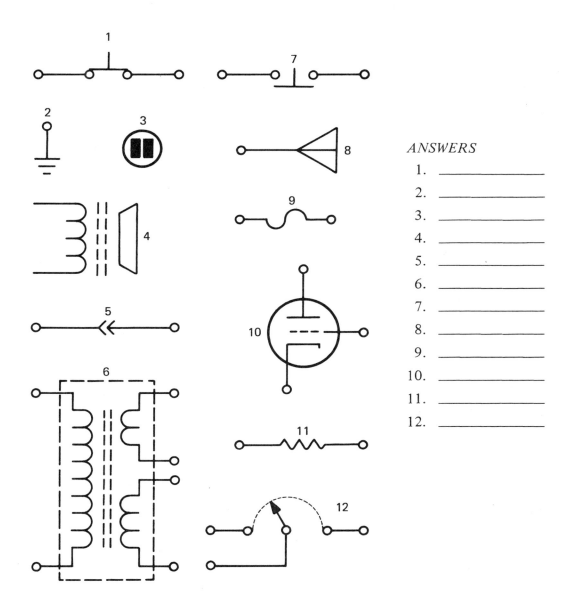

ANSWERS

1. _____

2. _____

3. _____

4. _____

5. _____

6. _____

7. _____

8. _____

9. _____

10. _____

11. _____

12. _____

RELATED TERMS

Give a brief definition of each term as progress is made through the unit.

Electricity _____

Electronics _____

Amperes _____

Voltage _____

Ohm _____

Ohm's law _____

Alternating current (AC) _____

Direct current (DC) _____

Series circuit _____

Parallel circuit _____

Wire connection_____

Battery_____

Ground_____

Fuse_____

Switch_____

SPST/DPDT_____

Resistor_____

Capacitor_____

Condenser_____

Polarity_____

Farad_____

Induction coil_____

Transformer_____

Laminations_____

Electron tube_____

Semiconductor_____

Diode_____

Transistor_____

Amplifier_____

Integrated circuit_____

ELECTRONICS

Electronics is a science. It takes many years of study to know and understand some phases of electronics. This unit is designed to give the reader a basic knowledge of what the various electrical components are, how they work, and what they do. This is done because electronic drafting requires a basic knowledge of electricity in order to make practical and useful electronic drawings.

Current

All matter is made up of atoms. These atoms have a nucleus and electrons that orbit around the nucleus. Electrons are negatively charged and the nucleus is positively charged. When a substance has an excess of electrons, it is negatively charged. When it has a deficiency of electrons, it is positively charged.

Current flows when electrons (–) move from a negatively-charged substance to a positively-charged substance. *Electricity* could be defined as current moving through conductors, such as wire or transformers. *Electronics* could be defined as current moving through devices, such as transistors, diodes, and vacuum tubes.

Current is measured in *amperes*. The opposing force acting against the flow of electrons is called *resistance,* which is measured in *ohms*. The pressure needed to force current through the resistance is called *voltage,* which is measured in *volts*.

There is a mathematical formula to determine the amount of voltage, resistance, or current. Figure 9-1 shows the symbol that is used with this mathematical formula known as *Ohm's law.* In this formula, E = voltage, I = amperage, and R = resistance.

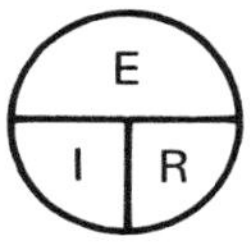

$$E = I \times R; I = \frac{E}{R}; R = \frac{E}{I}$$

Fig. 9-1 Ohm's law. Blocking out the unknown quantity gives the formula for finding it.

Direct Current

DC means *direct current.* Think of it as a straight "push" of voltage. It does not pulse as alternating current does. Direct current comes from a battery (chemical energy) or a direct-current generator, figure 9-2. Direct current can also be made by passing alternating current through a rectifier.

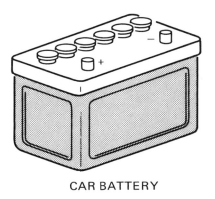

CAR BATTERY

Fig. 9-2 A car battery is a good example of direct current (one direction).

Alternating Current

Alternating current, AC, does exactly what its name implies, alternates directions. The standard alternating current used in the United States is 60 cycles per second. Figure 9-3 shows the wave form produced in one cycle of alternating current from a simple AC generator. It will repeat this cycle sixty times every second.

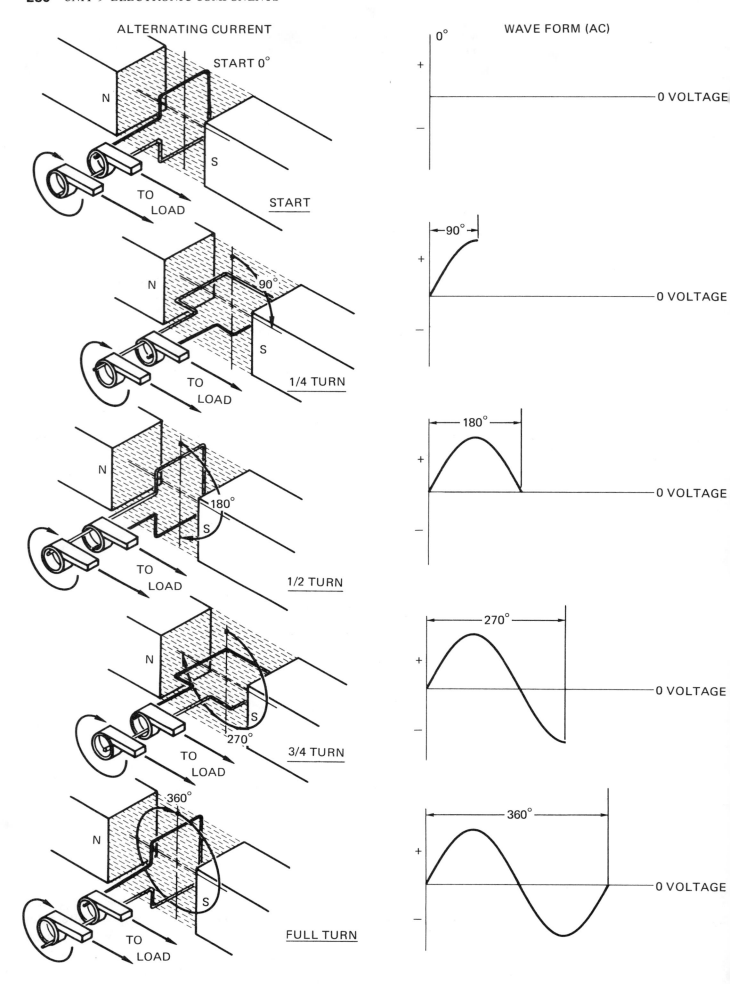

Fig. 9-3 One cycle of alternating current from a simple AC generator

SIMPLE CIRCUITS

When two unlike charges are connected by a *conductor* (wire), a current flows. In order to maintain current flow, an energy source must continue to add electrons to one end of the wire, causing a negative charge; and take electrons from the other end, causing a positive charge. If either end of the wire 's disconnected from the energy source, the charges in the wire become balanced and current stops flowing.

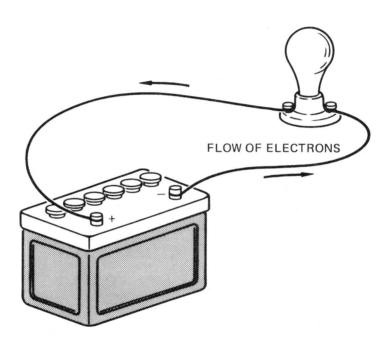

FLOW OF ELECTRONS

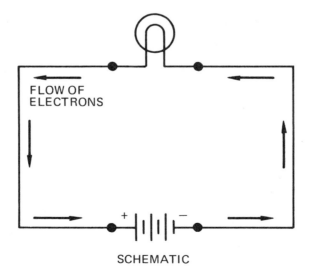

FLOW OF
ELECTRONS

SCHEMATIC

ELECTRONS FLOW FROM − TO +

Fig. 9-4 Current flow

Kinds of Circuits

There are three kinds of circuits: Series, parallel, and a combination of the two. Figure 9-5 shows series and parallel circuits.

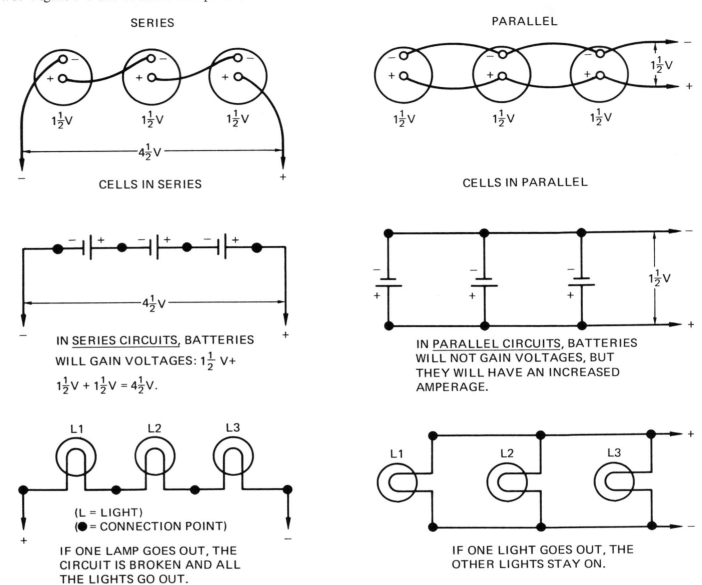

Fig. 9-5 Series and parallel circuits

Figure 9-6 shows a *combination* of both a series and a parallel circuit. L1 and L2 are in series with each other but are in parallel with L3 and L4. L5 and L6 are simply parallel circuits.

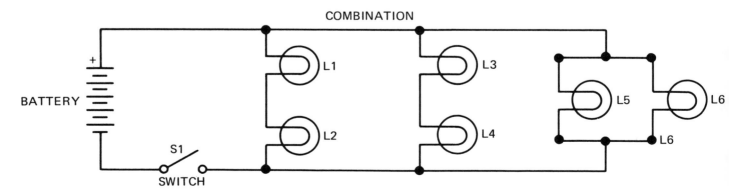

Fig. 9-6 Series-parallel circuit

SYMBOLS

It is important that a drafter memorize symbols and what function each represents. Symbols are drawn to scale using a standard electronic template.

Wires carry current to various parts. Many times wires cross over each other but do not connect. Other times wires are connected. Figure 9-7, left, illustrates wires that are connected; at the right are wires that do not connect. Use a 1/8-inch diameter circle template to make the arcs representing wires that do not connect and have them all go in the same direction. Darken in the circles showing connected wires.

Fig. 9-7 Wires

A battery is a form of power supply. It operates by changing chemical energy to electrical energy. A battery can have one cell, figure 9-8 at left; or more than one cell, figure 9-8 at right.

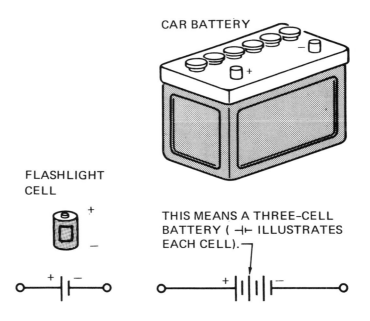

Fig. 9-8 Batteries or cells

A *ground* is referred to as a common point in a circuit. Grounds are all connected together, usually be metal chassis or frame, figure 9-9.

Fig. 9-9 Ground

A *fuse* is a safety device to limit the current in a circuit. If the limit is exceeded, the fuse burns out and turns the circuit off, figure 9-10.

Fig. 9-10 Fuse

SWITCHES

Switches are used in circuit to turn the electricity on and off. Study the basic switches in figure 9-11. Memorize how to draw them and how they work. SPST means single-pole, single-throw. Note what the other abbreviations mean.

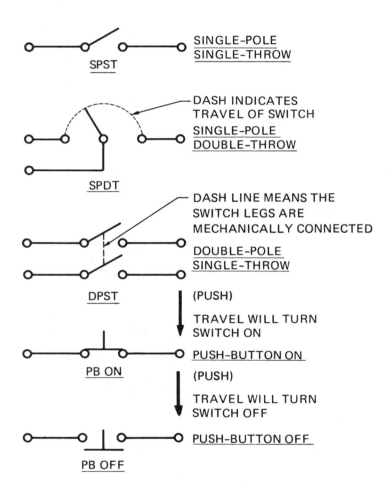

Fig. 9-11 Basic switches

Study the *wafer switch* in figure 9-12. In this position, current will travel from A to 3.

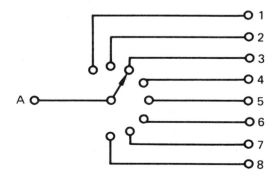

Fig. 9-12 Wafer switch

A *relay switch* is operated by an electromagnet, figure 9-13. Circuit 1 is turned on by switch 1 (S1) which magnetizes the *electromagnet*. This, in turn, pulls the switch down (note arrow) in switch 2 (S2), which turns on the circuit.

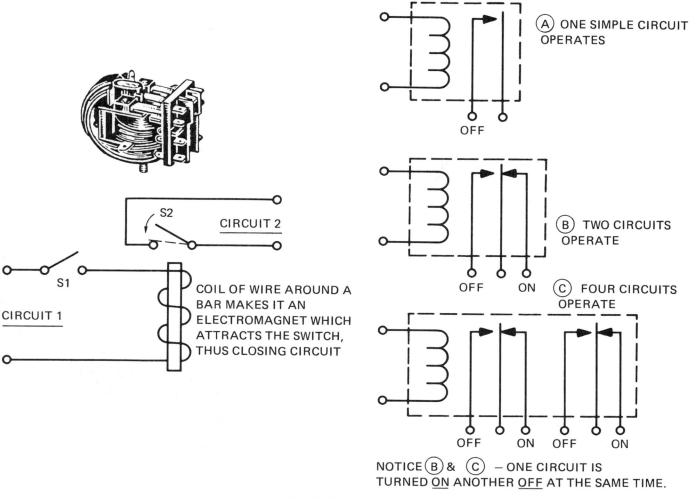

Fig. 9-13 Relay switch

Figure 9-14 illustrates various *industrial* switches: toggle, push-button, rocker, micro, wafer, interlocking, rotary.

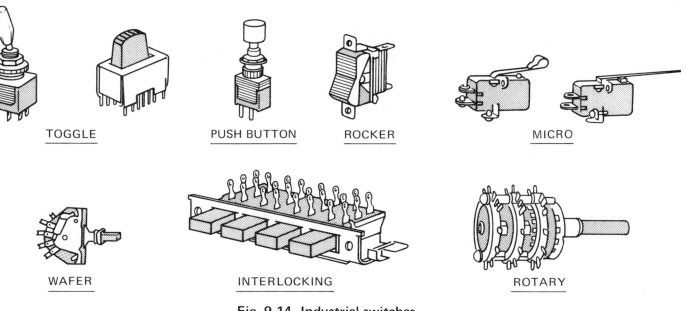

Fig. 9-14 Industrial switches

RESISTORS

Resistors reduce or regulate the current flow in a circuit. They also can be used to isolate one part of a circuit from another. A resistor converts electrical energy into heat energy. Usually, a resistor is made of carbon or a similar high electrical resistance material. Figure 9-15 shows four types of resistors.

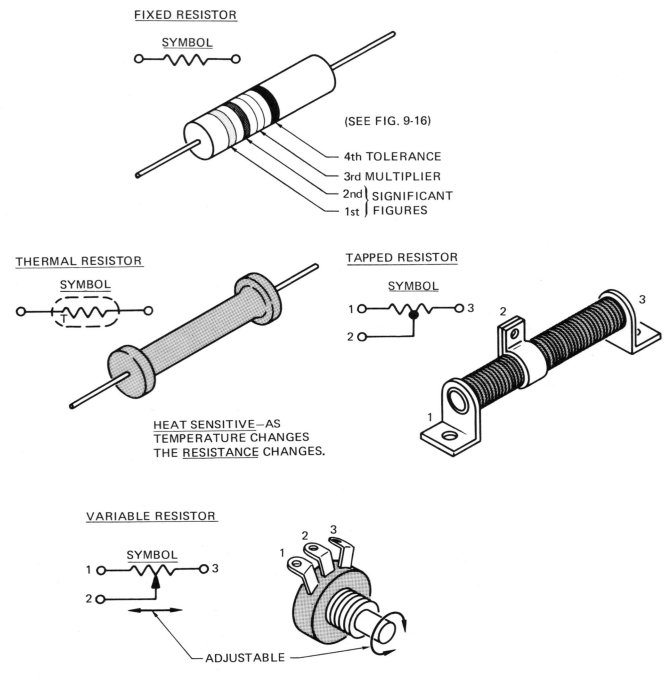

FIXED RESISTOR

SYMBOL

(SEE FIG. 9-16)

4th TOLERANCE
3rd MULTIPLIER
2nd } SIGNIFICANT
1st } FIGURES

THERMAL RESISTOR

SYMBOL

HEAT SENSITIVE—AS
TEMPERATURE CHANGES
THE RESISTANCE CHANGES.

TAPPED RESISTOR

SYMBOL

VARIABLE RESISTOR

SYMBOL

ADJUSTABLE

Fig. 9-15 Resistors

Resistors are color coded. Study the fixed resistor illustrated in figure 9-15. Note the four bands. The band closest to the end is counted as 1, the next is 2, etc. If a resistor had the following color numbers: orange, blue, red, and silver, what would be the value? Using the chart in figure 9-16: orange = 3, blue = 6, red = × 100, and silver = ± 10%. Thus, it would be: 3600 ± 10%. The 3, first digit; 6, second digit; times 100, third digit; equals 3600 plus or minus 10%.

OHM COLOR CODE CHART
Value in Ohms

Color Band	1st Band Digit 1	2nd Band Digit 2	3rd Band Multiplier X	4th Band Toler. %
Black	0	0	1	± 20%
Brown	1	1	10	± 1%
Red	2	2	100	± 2%
Orange	3	3	1,000	± 3%
Yellow	4	4	10,000	–
Green	5	5	100,000	± 5%
Blue	6	6	1,000,000	± 6%
Violet	7	7	–	± 12%
Gray	8	8	–	± 30%
White	9	9	–	± 10%
Gold	–	–	–	± 5%
Silver	–	–	–	± 10%
No Band	–	–	–	± 20%

Fig. 9-16 Resistor color code chart

CAPACITORS

Capacitors store or hold electrical energy, figure 9-17. By doing so, it smooths out ripples in a circuit. A capacitor also separates alternating from direct current as alternating current will pass through a capacitor but direct current will not. A capacitor is sometimes called a *condenser*.

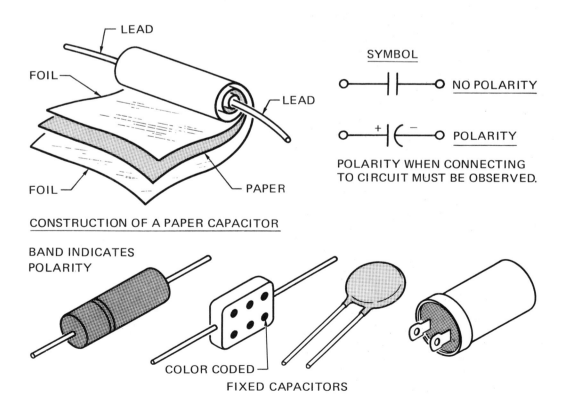

Fig. 9-17 Capacitors

Figure 9-18 illustrates what takes place in a capacitor. The water represents the electrical current. Some capacitors are *polarized*, or plus and minus. Polarity must be observed when connecting capacitors to circuits. The standard unit of measurement for a capacitor is the *farad* or *microfarad*.

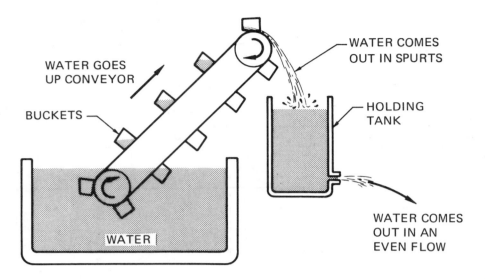

Fig. 9-18 Capacitor smooths out electrical current

The tuning control on a radio is an *adjustable capacitor*, figure 9-19.

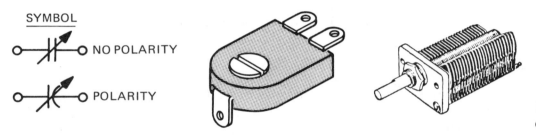

Fig. 9-19 Adjustable capacitor

INDUCTORS (COIL)

When a magnet is moved inside a coil of wire, the magnet's magnetic field cuts across the turns of wire in the coil. This action induces a voltage across the coil of wire. If the coil is a closed circuit, figure 9-20, it makes a current through the circuit. If the magnet is pulled up, the current goes one way; if it is pushed down, the current goes the other way.

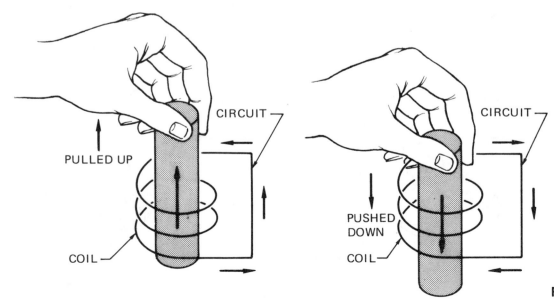

Fig. 9-20 Inductors

This relationship between magnetism and electricity is called *induction.* Thus, an *inductor* is simply a coil of wire. It can be coiled on a cardboard tube, called an air-core coil, or coiled on an iron core, called an iron-core coil, figure 9-21.

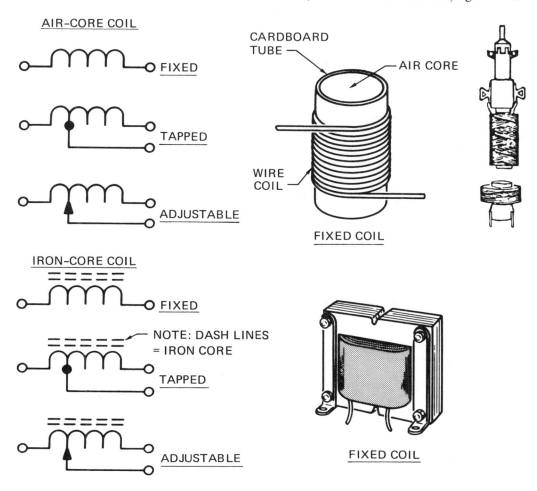

Fig. 9-21 Coils

TRANSFORMERS

The *transformer* is simply two or more inductance coils wound around a common core. It is used to change voltage in an alternating-current circuit. It will not work in a direct-current circuit. Input is put into the *primary winding* and output is taken from the *secondary winding(s),* figure 9-22.

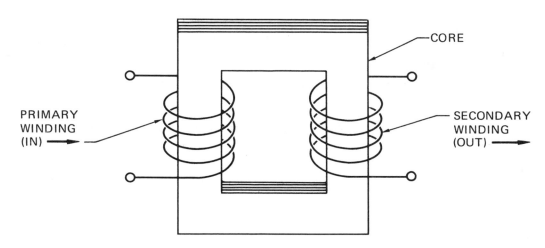

Fig. 9-22 Transformer

Transformer cores are usually made up of thin sheets of iron called *laminations*. These allow more magnetic lines between the coils so a greater amount of energy can be transferred from one coil to the other. The voltage induced across the secondary winding depends upon how many turns of wire it contains compared to the number of turns of wire that the primary winding contains. Thus 1000 turns primary winding with 500 turns secondary winding equals a step down of one half of the voltage, figure 9-23.

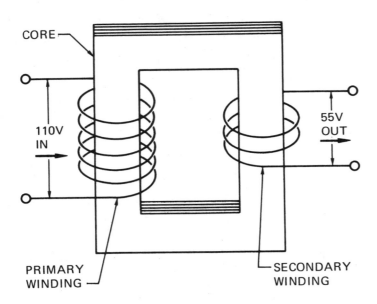

IF THE TRANSFORMER REDUCES VOLTAGE IT IS CALLED A <u>STEP-DOWN</u> TRANSFORMER.

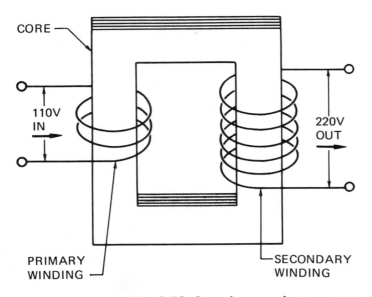

IF THE TRANSFORMER INCREASES VOLTAGE IT IS CALLED A <u>STEP-UP</u> TRANSFORMER.

Fig. 9-23 Step-down and step-up transformers

The transformer changes voltage in an alternating-current circuit up or down from the source coming to the transformer. Study the illustration in figure 9-24. The dash lines between coils means it is an iron-core transformer.

The relationship between the input voltage and the output voltage is in ratio to the coil turns on the primary and secondary coils. This ratio is written with E equalling voltage as:

$$\frac{\text{Input E} = \text{Turns in Primary}}{\text{Output E} = \text{Turns in Secondary}}$$

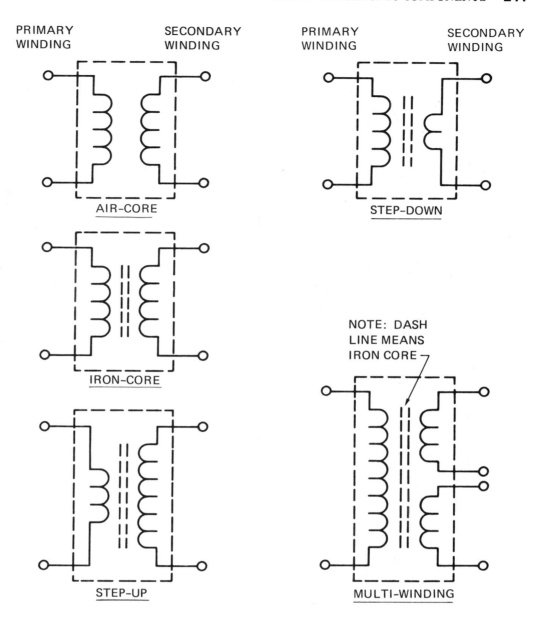

Fig. 9-24 Schematics of transformers

SEMICONDUCTORS

A *semiconductor* is a material that is not a good conductor of electricity or a good insulator of electricity. Sometimes it is a conductor, sometimes it is an insulator. Good semiconductors are germanium, silicon, and selenium. They are used to make transistors and diodes.

Diodes

Diodes allow electron flow in only one direction. To do this it uses a semiconductor material like germanium, figure 9-25.

Pure germanium is a good insulator, but if a little arsenic is added to it, it gives off electrons (–). This makes the germanium a negative or N-type material. If indium is added to pure germanium, it attracts the electrons which are negative. This germanium is a P-type (+) material.

Electrons flow *from* the germanium with arsenic side *to* the germanium with indium side. The electrons cannot flow the other way, thus current flows in only one direction. Diodes are used to convert alternating current to direct current.

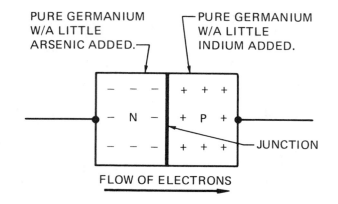

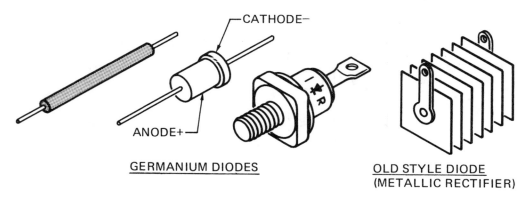

GERMANIUM DIODES

OLD STYLE DIODE
(METALLIC RECTIFIER)

NOTE: THE DIODE
REPLACES THE OLD
DIODE ELECTRON TUBE.

SYMBOL

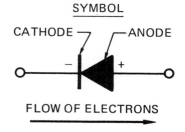

FLOW OF ELECTRONS

Fig. 9-25 Diodes

Transistors

The transistor is made up of the same materials as the solid state diode. In fact, the transistor is much like attaching two diodes together, back-to-back. A transistor has three elements: The base, the collector, and the emitter, figure 9-26. It works like a triode electron tube.

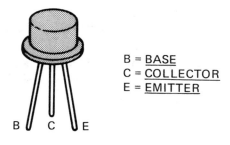

B = BASE
C = COLLECTOR
E = EMITTER

Fig. 9-26 Transistor

A very small voltage applied to the base can control a large voltage collector, thus it amplifies. There are two major kinds of transistors, the NPN type and the PNP type, figure 9-27.

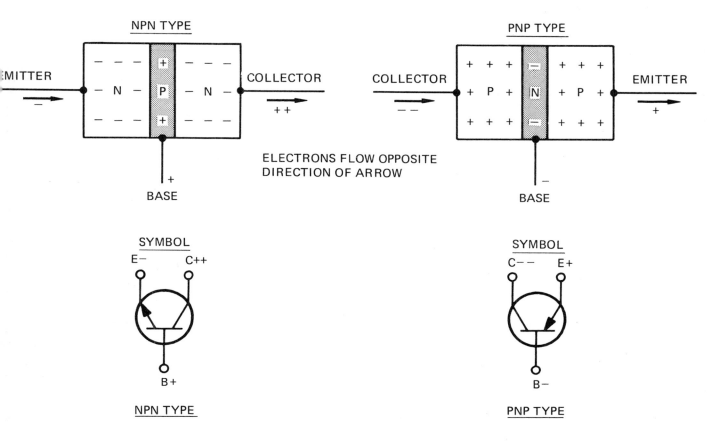

Fig. 9-27

POWER SUPPLY

The power supply is that part of a unit that converts 110 volts power to the direct-current power requirements. The power transformer (T1) converts the incoming 110 volts AC up or down to other voltages. Most transformers have more than one winding for various voltages. Figure 9-28 shows a very common use of the diode (D1, D2, D3, and D4) which *rectifies* the AC voltage to DC voltage. R1, C1, and C2 smooth out the pulsating direct-currect voltage to a smooth direct-current output.

FULL WAVE BRIDGE RECTIFIER

SAMPLE OF A STANDARD FULL
WAVE RECTIFIER.

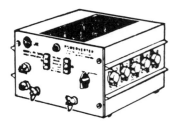

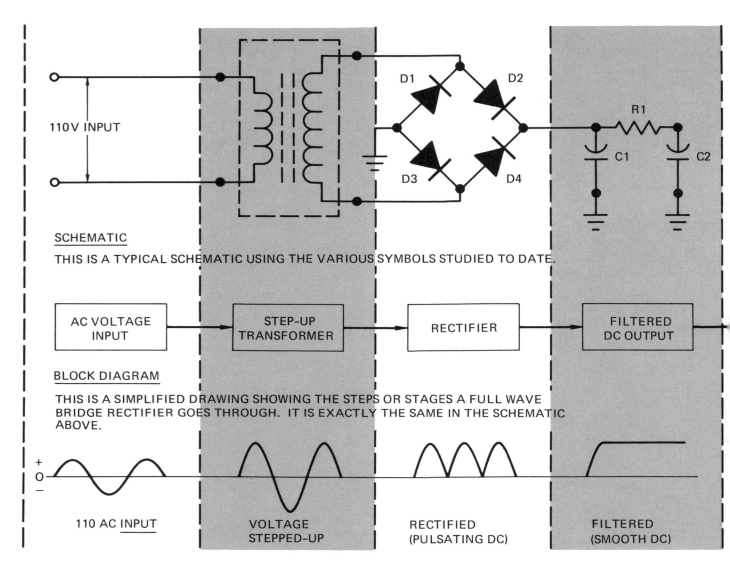

SCHEMATIC

THIS IS A TYPICAL SCHEMATIC USING THE VARIOUS SYMBOLS STUDIED TO DATE.

| AC VOLTAGE INPUT | STEP-UP TRANSFORMER | RECTIFIER | FILTERED DC OUTPUT |

BLOCK DIAGRAM

THIS IS A SIMPLIFIED DRAWING SHOWING THE STEPS OR STAGES A FULL WAVE
BRIDGE RECTIFIER GOES THROUGH. IT IS EXACTLY THE SAME IN THE SCHEMATIC
ABOVE.

110 AC INPUT   VOLTAGE STEPPED-UP   RECTIFIED (PULSATING DC)   FILTERED (SMOOTH DC)

WAVE FORM
THIS WILL GIVE YOU A VISUAL DESCRIPTION OF EXACTLY WHAT THE VOLTAGE IS DOING. IT AGREES WITH
THE TWO SKETCHES ABOVE. (GO FROM LEFT TO RIGHT.)

Fig. 9-28 Rectifier diagrams

ELECTRON TUBE

Diode and *triode tubes* are two of many kinds of tubes, figure 9-29. Other
tubes have more grids and plates in order to improve the control of electrons
and/or control them in other ways. Some are placed in a vacuum inside glass
envelopes.

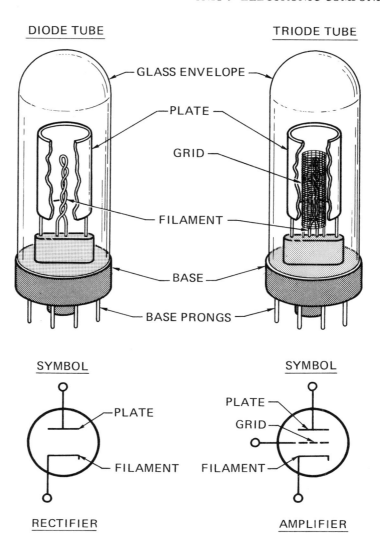

DIODE TUBE   TRIODE TUBE

GLASS ENVELOPE

PLATE

GRID

FILAMENT

BASE

BASE PRONGS

SYMBOL   SYMBOL

PLATE   PLATE

GRID

FILAMENT   FILAMENT

RECTIFIER   AMPLIFIER

Fig. 9-29 Diode and triode tubes

When tubes are drawn in a circuit, a base diagram is used along with a numbering system to identify the prongs, figure 9-30. All tubes illustrated are from the bottom of the tube. Tubes have their prongs arranged so they can only be put in one way. Seven and nine prong miniature tubes have the prongs molded into the glass shell with a space, as illustrated.

In order to know what prong number connects to inside the tube, a tube reference book must be used. Check the tube number against the reference book description. Prong #1 is found clockwise from either a space in the pins or the bottom of the key, as illustrated. Simply count clockwise from #1.

POWER AMPLIFIER

The power amplifier receives a low voltage output from the preamplifier and amplifies it enough so it can operate the speaker system. Figure 9-31 illustrates how the tube and transistor works in exactly the same way. The power signal is fed into the base of the transistor (grid of the tube), amplified, and comes out the collector of the transistor (plate of the tube).

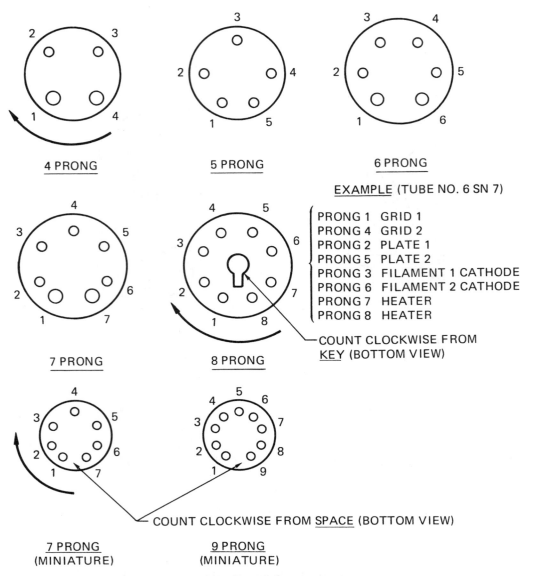

Fig. 9-30 Prong numbering system

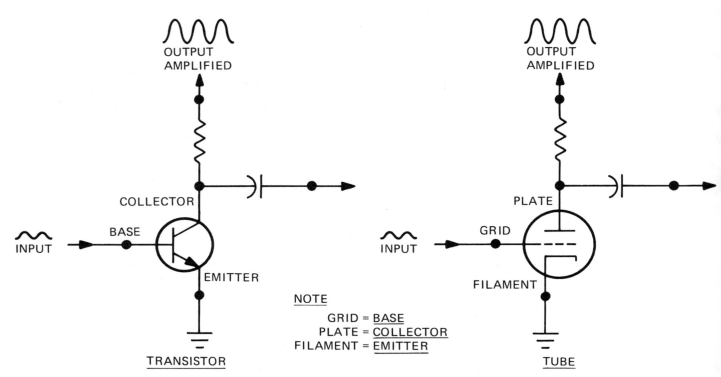

Fig. 9-31 Amplifier

INTEGRATED CIRCUIT

An *integrated circuit* has many components formed into it, such as transistors, diodes, capacitors, and resistors. These circuits are subminiature, approximately 1/16 inch square in size, which greatly reduces the physical size and weight of the assembly. Study the various sketches in figure 9-32. It gives a basic idea of what an integrated circuit is. It is often mounted in a complete unit. This complete unit is large only because of the physical problem of making external connections. The schematic illustrates the components a circuit may contain.

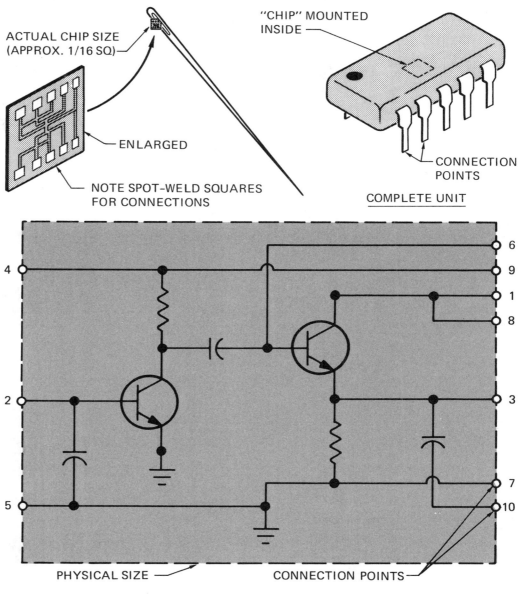

COMPLETE UNIT

Fig. 9-32 Integrated circuitry

DIMENSIONED PLANS (SPECIFICATIONS) OF COMPONENTS

In designing various assemblies or subassemblies, the drafter must use standard manufactured components. Manufacturers of these components provide specification sheets. These sheets give the general appearance, capabilities, overall dimensions, and mounting hole locations.

Figures 9-33, 9-34, and 9-35 show three such specification plans.

AC AMMETERS

Range	Ohms Res.	Catalog Numbers for Model 230-G	231-S	237-S
0-1	.38	153-0001	153-0348	153-0377
0-2	.092	153-0003	153-0349	153-0378
0-3	.039	153-0005	153-0350	153-0379
0-5	.014	153-0006	153-0351	153-0380
0-10	.004	153-0065	153-0352	153-0381
0-15	.003	153-0066	153-0353	153-0382
0-25	.002	153-0068	153-0354	153-0383
0-30	.001	153-0069	153-0355	153-0384
0-50	.00035	153-0070	153-0356	153-0385

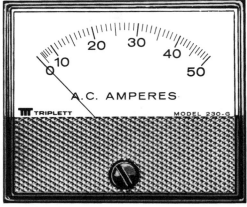

230-G

DC MILLIVOLTMETERS (Pivot & Jewel)

Range	Ohms Res.	Catalog Numbers for Model 220-G	221-T	227-T
0-50	6.7	152-0183	152-1451	152-1520

DC VOLTMETERS

Range	Ohms/ Volt	Catalog Numbers for Model 220-G	221-T	227-T
0-5	1000	152-0193	152-1456	152-1525
0-10	1000	152-0195	152-1457	152-1526
0-25	1000	152-0198	152-1459	152-1528
0-30	1000	152-0199	152-1460	152-1529
0-50	1000	152-0200	152-1461	152-1530
0-100	1000	152-0202	152-1462	152-1531
0-150	1000	152-0203	152-1463	152-1532

231-S

AC VOLTMETERS

Range	Ohms/Volt	Catalog Numbers for Model 230-G	231-S	237-S
0-10	10	153-0072	153-0360	153-0389
0-50	50	153-0077	153-0363	153-0392
0-100	91	153-0022	153-0364	153-0393
0-150	125	153-0023	153-0365	153-0394
0-250	144	153-0025	153-0366	153-0395
0-300	144	153-0026	153-0367	153-0396
0-500 *	125	153-0214	153-0368	153-0397

237-S

RF THERMOAMMETERS

Range	Ohms Res.	Catalog Numbers for Model 240-G	241-T	247-T
0-1	.35	52-3435	152-1489	152-1557
0-1.5	.21	52-3531	152-1490	152-1558
0-5	.06	52-3169	152-1492	152-1560
0-10	.03	52-3422	152-1493	152-1561

Fig. 9-33 Specifications of components

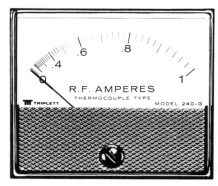

240-G

G SERIES • BEZEL & ILLUMINATION

Illumination Kit Requires Bezel Mounting

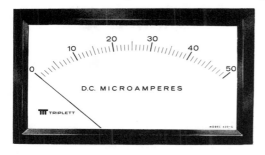

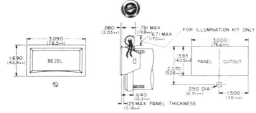

Allow .50" additional width for Bezel mounting hardware

2½" MODELS
Model 220-G Series
Metal Bezel—Cat. No. 13-238
Illumination Kit—Cat. No. 66-80

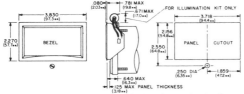

Allow .50" additional width for Bezel mounting hardware

3½" MODELS
Model 320-G Series
Metal Bezel—Cat. No. 13-239
Illumination Kit—Cat. No. 66-80

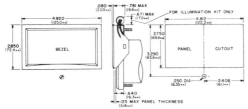

Allow .50" additional width for Bezel mounting hardware

4½" MODELS
Model 420-G Series
Metal Bezel—Cat. No. 13-236
Illumination Kit—Cat. No. 66-80

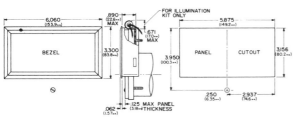

Allow .50" additional width for Bezel mounting hardware

5½" MODELS
Model 520-G Series
Metal Bezel—Cat. No. 13-240
Illumination Kit—Cat. No. 66-83

SHALLOW BARREL   G SERIES

A special shallow barrel, GS, Thermoplastic back is available for 1½", 2½", 3½" and 4½" G-Series panel instruments.

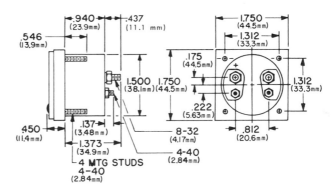

120-GS

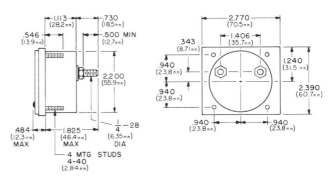

220-GS

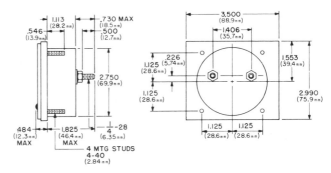

320-GS

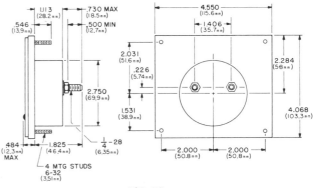

420-GS

Fig. 9-34

ENGR. FORM 71-103 K&E 10 5156 11-65•

MODEL 320-

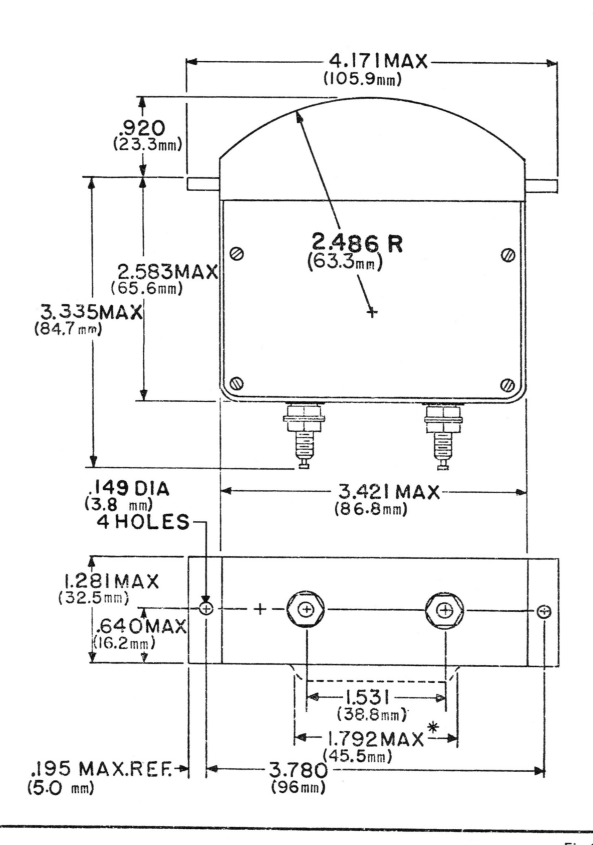

Fig. 9-35

& 330-E

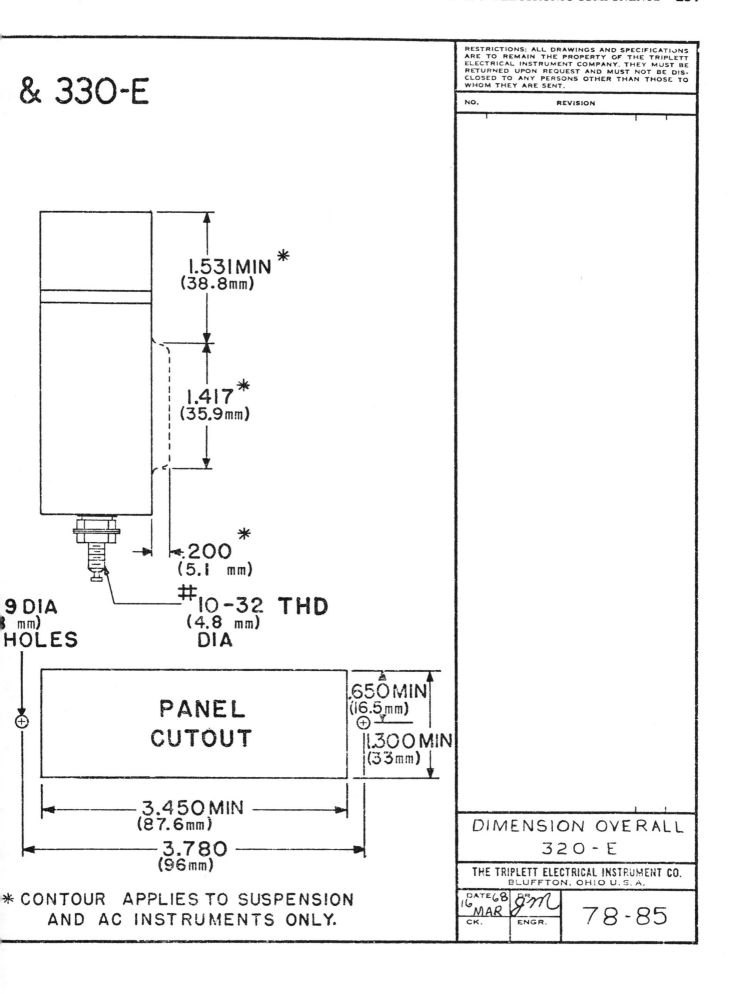

1.531 MIN *
(38.8mm)

1.417 *
(35.9mm)

.200 *
(5.1 mm)

#10-32 THD
(4.8 mm)
DIA

9 DIA
(mm)
HOLES

PANEL
CUTOUT

.650 MIN
(16.5mm)

1.300 MIN
(33mm)

3.450 MIN
(87.6mm)

3.780
(96mm)

*** CONTOUR APPLIES TO SUSPENSION
AND AC INSTRUMENTS ONLY.**

NO.	REVISION		

DIMENSION OVERALL
320 - E

THE TRIPLETT ELECTRICAL INSTRUMENT CO.
BLUFFTON, OHIO U.S.A.

DATE 68	BR	
16 MAR		78-85
CK.	ENGR.	

UNIT REVIEW

One-hour time limit

Questions

1. Explain the difference between alternating current and direct current. How is each produced?

2. Explain which direction current flows and which direction electrons flow.

3. What is a fuse?

4. What does DPST mean?

5. What is a good example of a variable resistor?

6. If there are 1000 turns in the primary winding of a transformer with 110 volts coming into it, and 500 turns on the secondary winding, what is its output?

7. What does a transistor do?

8. What is the simple function of a capacitor?

9. Explain how a diode tube works.

10. How do series and parallel circuits differ?

Identification

Identify what each symbol represents and place its name in the space provided.

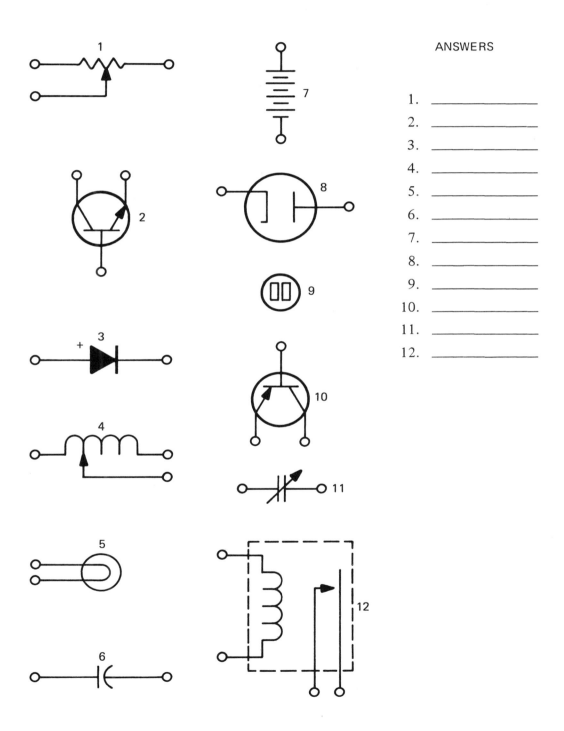

ANSWERS

1. _____
2. _____
3. _____
4. _____
5. _____
6. _____
7. _____
8. _____
9. _____
10. _____
11. _____
12. _____

Before proceeding to the next unit:

_____ Instructor's approval

_____ Progress plotted

UNIT 10

ELECTRONIC DRAWING

OBJECTIVE

The student will learn to make basic electronic drawings from rough sketches.

PRETEST

Two-hour time limit

Questions

1. What is a block diagram?

2. On a harness layout, what does a note like BL-6 mean?

3. What kind of an electronic drawing illustrates a circuit showing exactly how each component is connected (point-to-point connection, not as they will actually be located in the unit)?

4. What does a power supply do? Explain in full detail.

5. What is a printed circuit?

6. List four kinds of electronic drawings.

7. Why should a grid be used in laying out a schematic drawing?

8. Explain line weight as it applies to electronic drawing.

9. Which kind of drawing illustrates the parts as they actually look?

Problem

From the rough sketch, make a complete "finished" schematic drawing on a 3/8-inch grid. Label each component as illustrated. Use standard electronic symbols.

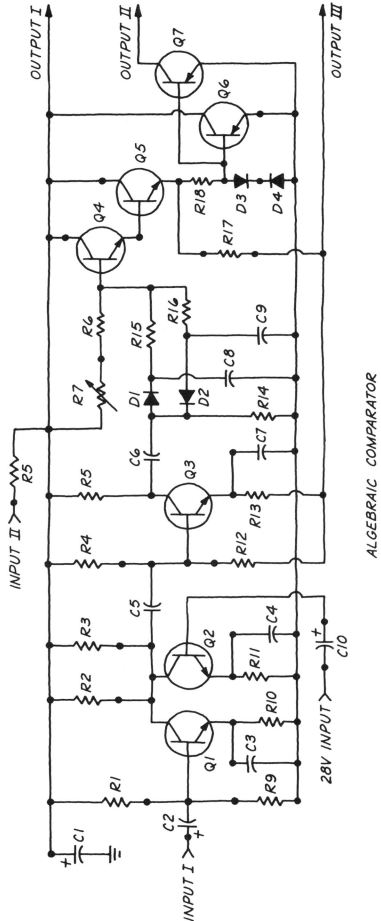

RELATED TERMS

Give a brief definition of each term as progress is made through the unit.

Electrical template _____

Parts list drawing _____

Block diagram _____

Harness layout _____

Pictorial drawing _____

Schematic drawing _____

Printed circuits _____

Resist paint _____

Point-to-point contact _____

Pad _____

Conductor _____

Drilling drawing _____

Circuit description _____

LINE WORK IN ELECTRONIC DRAWINGS

1. Lines representing wires must be black, medium weight, and have sharp corners.
2. Dash lines representing subassemblies or large components must be black and center-line weight.
3. Lay out a light grid before starting. Evenly space neat lines 1/4 inch (6) apart over the grid to represent wires.
4. Design line work with the least amount of crossovers.
5. Long parallel lines should be arranged in groups of three lines or less. More than three are difficult to follow.
6. Leave space around all component symbols and their connection points for the callouts.
7. Make all lines neat and consistent. Make all resistors the same size, etc. There are various electrical templates on the market, such as the one shown in figure 10-1.

PARTS LIST DRAWING OR PHOTOGRAPH

A *parts list drawing* or photgraph shows exactly what a complete assembly looks like and approximately where each subassembly is located. Study the parts list photos, figures 10-2 and 10-3, illustrating both the inside and outside of an assembly.

A close-up of the solo voicing and preamp circuit identifying each part, figure 10-4, illustrates where each part is located and notes its part number.

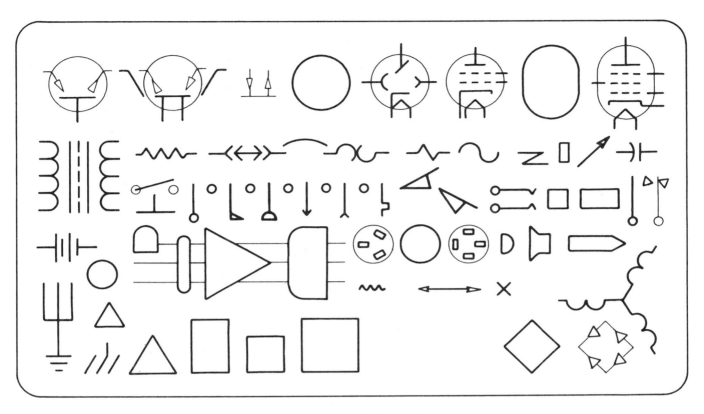

Fig. 10-1 Electrical template

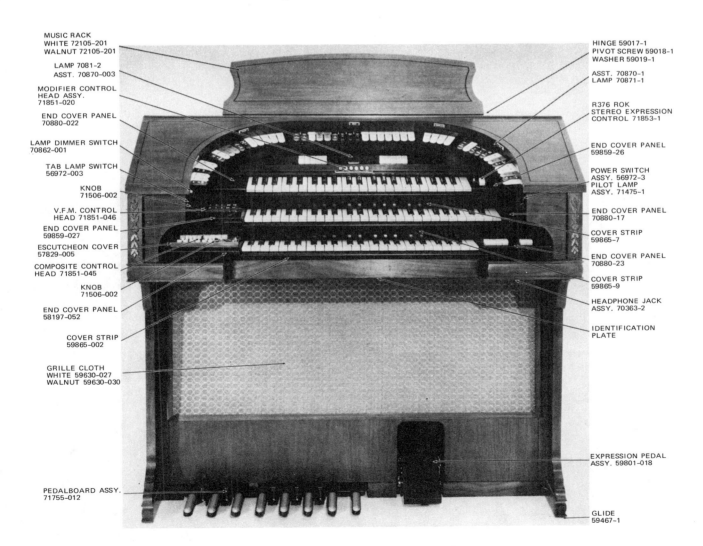

MUSIC RACK
WHITE 72105-201
WALNUT 72105-201

LAMP 7081-2
ASST. 70870-003

MODIFIER CONTROL
HEAD ASSY.
71851-020

END COVER PANEL
70880-022

LAMP DIMMER SWITCH
70862-001

TAB LAMP SWITCH
56972-003

KNOB
71506-002

V.F.M. CONTROL
HEAD 71851-046

END COVER PANEL
59859-027

ESCUTCHEON COVER
57829-005

COMPOSITE CONTROL
HEAD 71851-045

KNOB
71506-002

END COVER PANEL
58197-052

COVER STRIP
59865-002

GRILLE CLOTH
WHITE 59630-027
WALNUT 59630-030

PEDALBOARD ASSY.
71755-012

HINGE 59017-1
PIVOT SCREW 59018-1
WASHER 59019-1

ASST. 70870-1
LAMP 70871-1

R376 ROK
STEREO EXPRESSION
CONTROL 71853-1

END COVER PANEL
59859-26

POWER SWITCH
ASSY. 56972-3
PILOT LAMP
ASSY. 71475-1

END COVER PANEL
70880-17

COVER STRIP
59865-7

END COVER PANEL
70880-23

COVER STRIP
59865-9

HEADPHONE JACK
ASSY. 70363-2

IDENTIFICATION
PLATE

EXPRESSION PEDAL
ASSY. 59801-018

GLIDE
59467-1

Fig. 10-2 Parts list photo — outside view

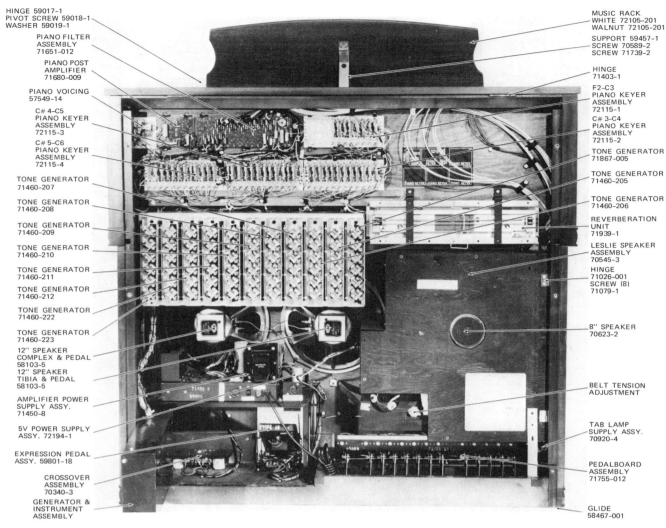

HINGE 59017-1
PIVOT SCREW 59018-1
WASHER 59019-1
PIANO FILTER
ASSEMBLY
71651-012
PIANO POST
AMPLIFIER
71680-009
PIANO VOICING
57549-14
C# 4-C5
PIANO KEYER
ASSEMBLY
72115-3
C# 5-C6
PIANO KEYER
ASSEMBLY
72115-4
TONE GENERATOR
71460-207
TONE GENERATOR
71460-208
TONE GENERATOR
71460-209
TONE GENERATOR
71460-210
TONE GENERATOR
71460-211
TONE GENERATOR
71460-212
TONE GENERATOR
71460-222
TONE GENERATOR
71460-223
12" SPEAKER
COMPLEX & PEDAL
58103-5
12" SPEAKER
TIBIA & PEDAL
58103-5
AMPLIFIER POWER
SUPPLY ASSY.
71450-8
5V POWER SUPPLY
ASSY. 72194-1
EXPRESSION PEDAL
ASSY. 59801-18
CROSSOVER
ASSEMBLY
70340-3
GENERATOR &
INSTRUMENT
ASSEMBLY

MUSIC RACK
WHITE 72105-201
WALNUT 72105-201
SUPPORT 59457-1
SCREW 70589-2
SCREW 71739-2
HINGE
71403-1
F2-C3
PIANO KEYER
ASSEMBLY
72115-1
C# 3-C4
PIANO KEYER
ASSEMBLY
72115-2
TONE GENERATOR
71867-005
TONE GENERATOR
71460-205
TONE GENERATOR
71460-206
REVERBERATION
UNIT
71939-1
LESLIE SPEAKER
ASSEMBLY
70545-3
HINGE
71026-001
SCREW (8)
71079-1
8" SPEAKER
70623-2
BELT TENSION
ADJUSTMENT
TAB LAMP
SUPPLY ASSY.
70920-4
PEDALBOARD
ASSEMBLY
71755-012
GLIDE
58467-001

Fig. 10-3 Parts list photo — inside view

Q286 71226
C286 59622
SOLO CIRCUIT
SOLO CIRCUIT
SOLO VOICE
C289 59622
Q287 71226
Q314 71226
SOLO PREAMP
C297 7149
SOLO CIRCUIT
SOLO VOICE
SOLO CIRCUIT
Q288 71226
SOLO CIRCUIT
C298 596
SOLO CIRCUIT

Fig. 10-4 Parts list photo — close-up view

BLOCK DIAGRAMS

Block diagrams use rectangles or blocks to represent a group of components that do a complete function, such as a power supply or an amplifier. This system illustrates a very complex piece of equipment and the relationship between the subassemblies. The block diagram reads like a page of a book. Input is at the top left, goes from left to right, with the output at the bottom right, figure 10-5.

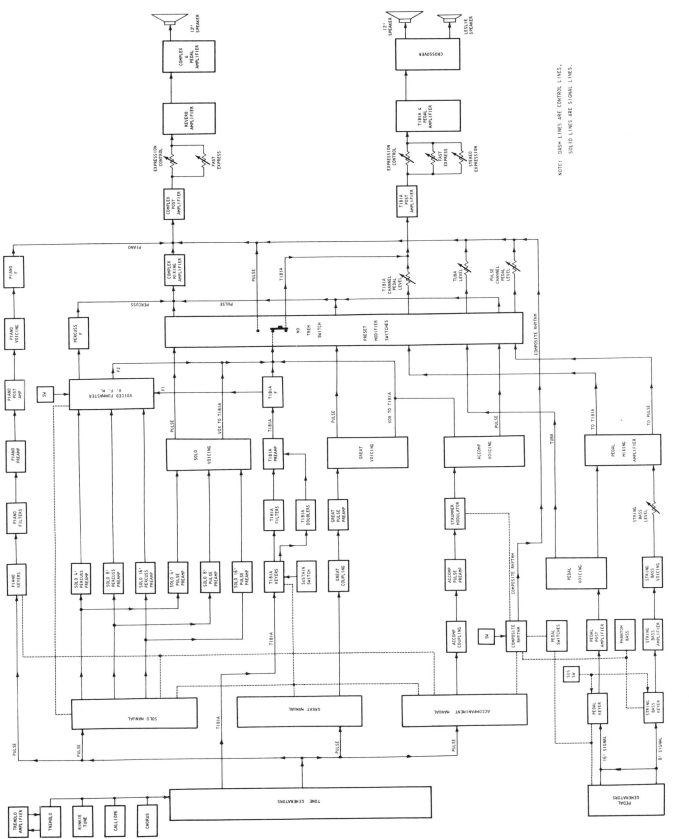

Fig. 10-5 Block diagram

HARNESS LAYOUT

Harnesses are usually completely made up and cut to correct lengths in what is called a subassembly. Usually, ends are marked or tagged. Notice end 5 at the far left of figure 10-6. The top wire is GR1. This means the green wire goes to terminal number one. Study the sample of terminal number four.

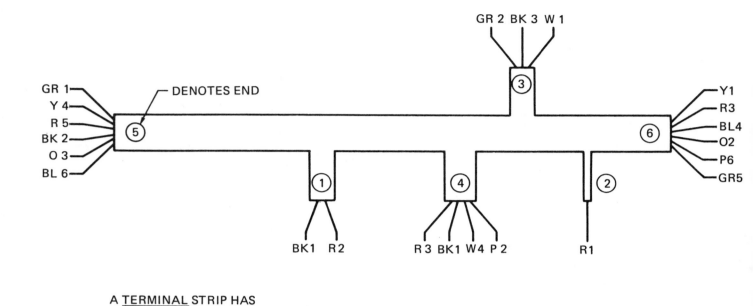

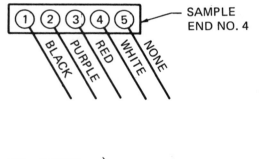

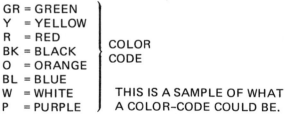

GR = GREEN
Y = YELLOW
R = RED
BK = BLACK   COLOR
O = ORANGE   CODE
BL = BLUE
W = WHITE   THIS IS A SAMPLE OF WHAT
P = PURPLE   A COLOR-CODE COULD BE.

Fig. 10-6 Harness layout

KINDS OF DRAWINGS

There are various kinds of electronic drawings. Use the kind that best describes what you are designing. Study the *pictorial drawing* and *schematic diagram*, figures 10-7 and 10-8. Both represent the same simple radio.

Pictorial drawings show the circuit as it looks to the eye and are used by the worker who puts the unit together, figure 10-7.

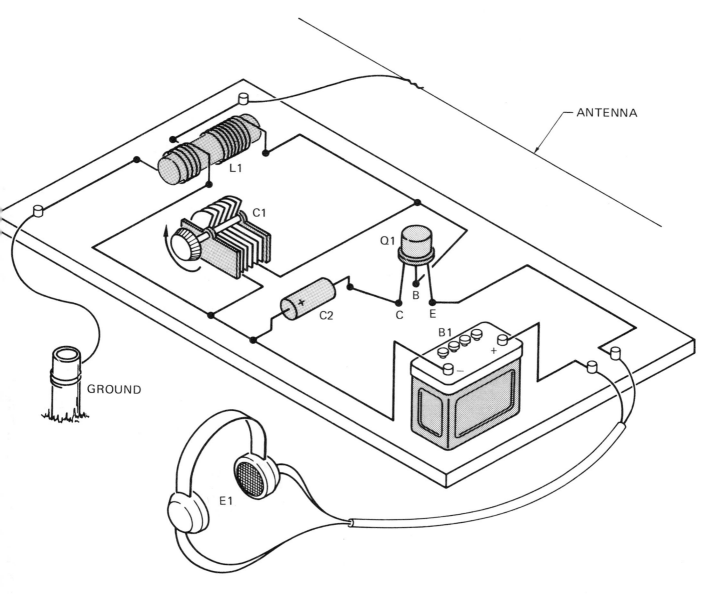

Fig. 10-7 Pictorial drawing of a radio

Schematic drawings use symbols to show each part and exactly how each part is connected in the circuit. A drafter must know the symbols in order to read a schematic diagram, figure 10-8.

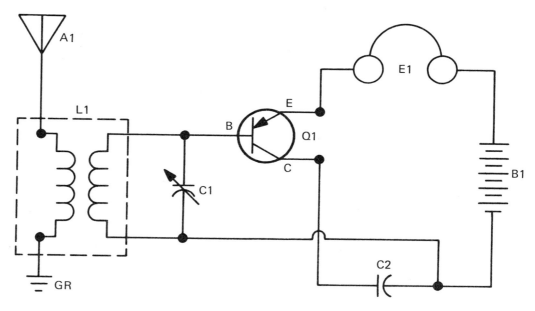

Fig. 10-8 Schematic diagram of the same radio shown in figure 10-8

Drawing a Schematic

Figure 10-9 is a finished drawing of a power supply. To draw it, follow steps 1 through 4.

1. Make a very light 3/8-inch (10) grid on another sheet of drawing paper using a sharp 4H pencil.

2. Lightly sketch the power supply on grid lines as shown in figure 10-10. Be sure to stay either on the grid lines or in the center between the grid lines as shown.

3. Stop and check all work after completing the sketch on the grid. Check all connections, spacing, and overall neatness.

4. Now trace the sketch and complete the drawing. Watch all line weights. Use an electrical template to draw the component symbols.

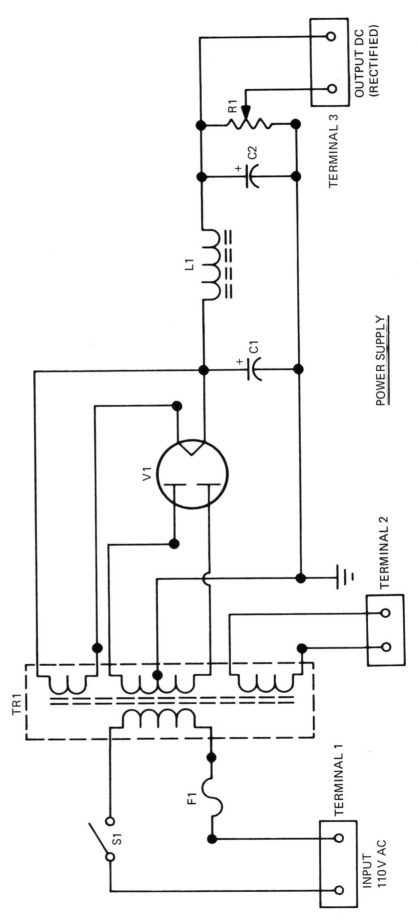

Fig. 10-9 Finished schematic of a power supply

NOTE: 1/4 inch (6) is minimum for grid size. The "V" is used for tube callouts. The "L" for coils. Number each part from left to right. The input is always on the left and the output on the right. Your sketch should look exactly like this.

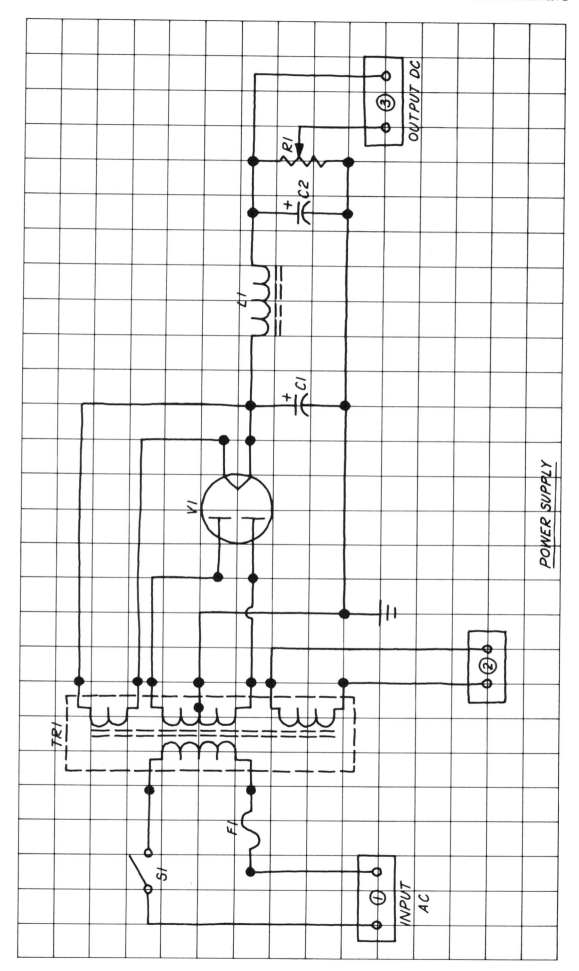

Fig. 10-10 Sketch of power supply on 3/8-inch grid

Practice Exercise 10-1

Make a finished drawing of this sketch on a 3/8-inch (10) grid. Use correct line weight, standard symbols, and drawing procedures. Call out each component from left to right. Be sure the grid does not show on the final copy. Place the input on the left and the output on the right. Try to trace the input signal from the antenna through tube VI (amplifier) to output terminal 5. Note that the AC power from terminal 2 supplies filament of tube. Compare your work to the answer on page 283.

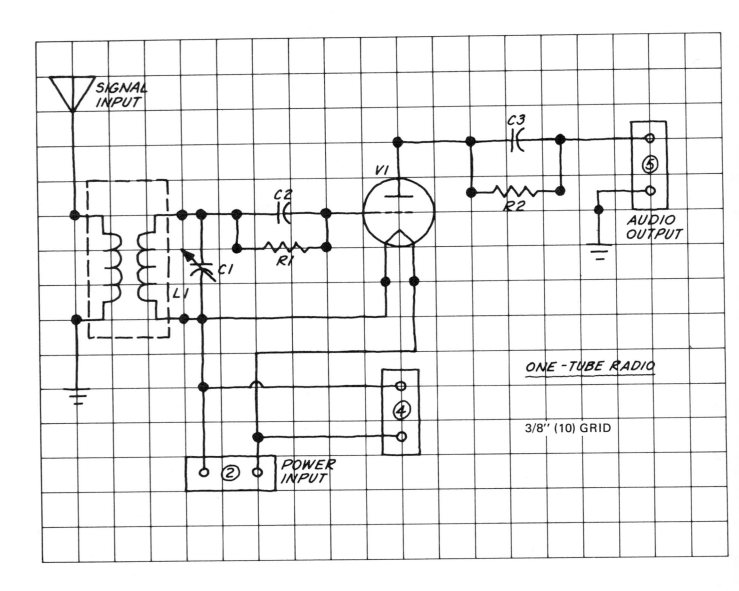

Practice Exercise 10-2

Make a finished drawing of the sketch on a 3/8-inch (10) grid. Use correct line weight, standard symbols, and drawing procedures. Call out each component from left to right. Compare your work to the answer on page 284.

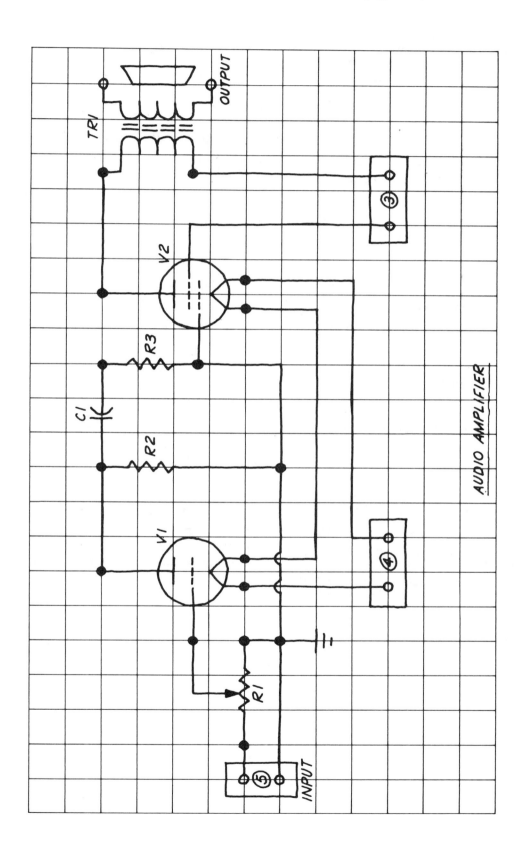

Practice Exercise 10-3

The power supply, one-tube radio, and audio amplifier drawn so far are placed left to right in a block drawing.

The next exercise is to draw a schematic of a complete radio following these steps:

1. Combine the three schematics completed in figure 10-10, exercise 10-1, and exercise 10-2 into one complete schematic of a radio.

2. Use a 3/8-inch grid on separate drawing paper. The schematic should be about 10″ × 16″.

3. Follow all the steps outlined in the other schematic drawings:
   a. Draw a light grid.
   b. Lightly sketch all lines and components on the grid.
   c. Check connections, spacings, and neatness.
   d. Trace the sketch.
   e. Check accuracy.

4. Label all components from left to right. This is a new drawing so start with number 1, left side, and call out numbers to the right side.

Compare your work to the answer on pages 284 and 285. The answer uses a 1/4-inch grid, but the drawing should be the same except for the size. Note that in order to combine all three drawings with the least amount of crossovers, some components must be rearranged. Add all terminals and label each. Be sure all point-to-point connections are the same.

Practice Exercise 10-4

Using a 1/2-inch (12) square grid, draw the rectifier power supply as a standard schematic drawing. Use correct symbols, line weight, and label each component. Check all work. Be sure point-to-point connections are correct. Input should be on the left and output on the right. Compare your work to the answer on page 286.

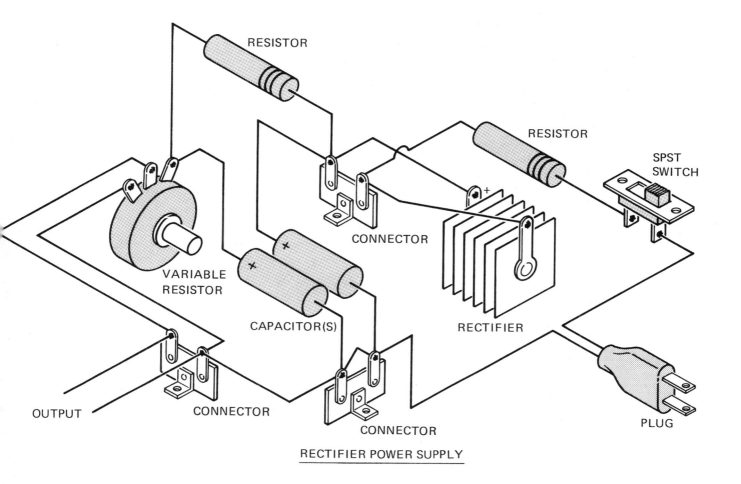

RECTIFIER POWER SUPPLY

Practice Exercise 10-5

Using a 1/2-inch (12) grid, draw the two-transistor radio as a standard schematic drawing. Use correct symbols, line weight, and label each component. Check all work. Place the input on the left and the output on the right. Compare your work to the answer on page 287.

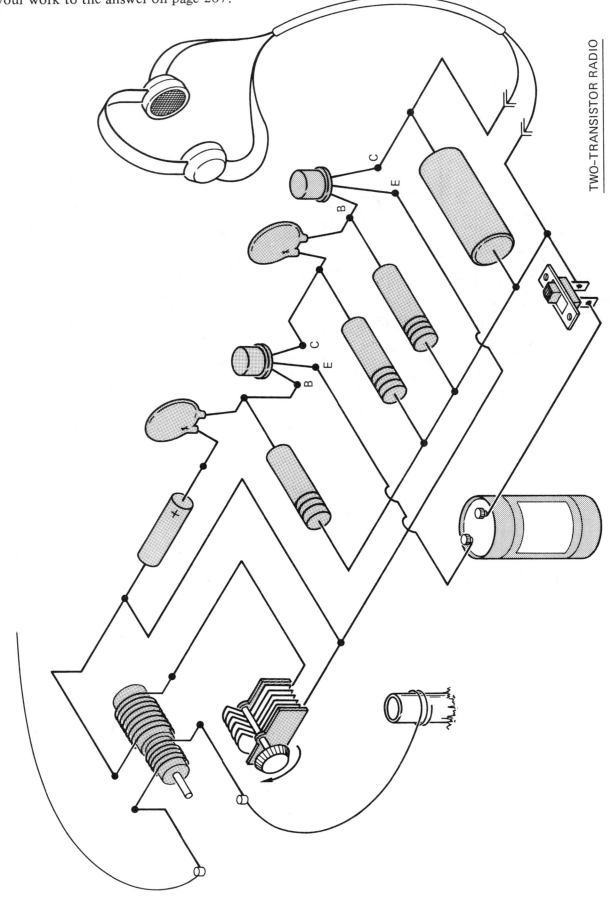

TWO-TRANSISTOR RADIO

Practice Exercise 10-6

Start with a 1/2-inch (12) grid. Make a quick sketch as outlined. Check all point-to-point contacts. If correct, trace neatly using all correct symbols. Watch line weight, neatness, and label each component as noted. Compare your work to the answer on page 288.

This schematic has three circuits tied together by a step down, multi-output transformer (T1).

CIRCUIT I

Starting from a standard male plug (P1) with a 110 volt AC input, design a circuit that has a pushbutton on switch (S1); a pushbutton off switch (S2); a fuse (F1); and a multioutput, step down transformer (T1) with a 110 V primary winding and 60 V secondary winding, 10 V secondary winding.

CIRCUIT II

Starting from the 10 V side of T1, draw a circuit that has an SPST switch (S3), a fuse (F2), and an SPST relay switch (S4).

CIRCUIT III

Starting from the 60 V side of T1, draw a circuit that is turned on/off by relay switch S4 which controls three lights in series (L1, L2, L3) and two lights in parallel (L4, L5). Note that L1, L2, L3, L4, and L5 equal the full load on this circuit. Be sure to fuse this circuit (F3). To meter this circuit, add a built-in amp meter (amp meters are connected in series with the load) and a built-in voltmeter (voltmeters are connected in parallel with the load).

POWER-SUPPLY AMPLIFIER

Figures 10-11 and 10-12 show an actual *circuit description* and a photo of the underside of that circuit. Figure 10-13 shows a schematic of the same circuit. Try to locate various components in the photograph from the schematic. As an example, locate resistor 901 (check point 5) and transistor Q902 (check point 6).

POWER AMPLIFIER AND POWER SUPPLY DESCRIPTION

The Power Amplifier and Power Supply are on one chassis, shown schematically. The power amplifier is a transistorized two channel amplifier, featuring three stages and 40 watts per channel. The output circuit is a single "ended" push pull or bridge circuit, coupled to the loud speaker. This DC coupled output circuit requires the balanced positive and negative power supply mentioned above, since the return signal current from the speaker flows through the power supply filter capacitors. Hum is cancelled in the output circuit and matched transistors. Transistors of the same color group must be used in pairs.

Tibia and pedal signals and complex channel signals are brought to the amplifier chassis through shielded cables and are coupled to the first stage (Q901 and Q905), or predriver. This is direct coupled to an emitter follower power stage (Q902 and Q906), which is in turn shunt coupled to the driver transformer and so to the output transistors. There are no controls associated with the amplifier power supply chassis, since bias voltages are fixed and level adjustments are made at the post amplifier board assembly.

Power supply for the two power amplifier stages and driver stages is an unfused +30 volts and –30 volts and wires directly from the full wave rectifier-filter sections. The transformer provides a full wave filtered + 80 volt keying voltage supply for fast attack tibia keyers and a full wave +40 volt filtered source for the +30 volt feedback regulator, Q910. The –30 volt and +30 volt regulated volt sources available for external use at pin 8 and pin 2 of the power plug P2, feed through 20 ohm internal resistors to limit external power demand to a safe level. Internal solid state short circuit protection (Q911) of the main. +30 volt source eliminates blown fuses caused by accidental momentary short circuits occurring when servicing the instrument.

The +30 volt regulated supply consists of a standard amplifier regulator. The output stage Q910 drives its load as an emitter follower, with the output feeding back to amplifier Q912 base via voltage divider R939 and R940. The Zener diode in the Q912 emitter circuit substitutes for a low value emitter resistor (and reference voltage), so that the amplifier has high effective gain. The series transistor, Q910, is turned on by current through R938. The regulator output voltage rises, is fed to Q912 forward biasing it, reducing the voltage drop across R938 and so the bias current available to Q910. Current flow through Q910 is reduced, dropping the regulator output voltage at pin 7, thus the voltage supply remains constant under varying conditions.

Short circuit protection is provided by the 1.3 ohm monitoring resistor R943 and associated transistor, Q911. With normal current drain through power plug pin 7 circuits, Q911 does not conduct. When current through R943 reaches around 500 MA, the voltage difference between Q911 base and emitter is sufficient to start current flow, causing Q910 base voltage to drop, reducing current through it and holding output power at a safe level.

Fig. 10-11 Circuit description

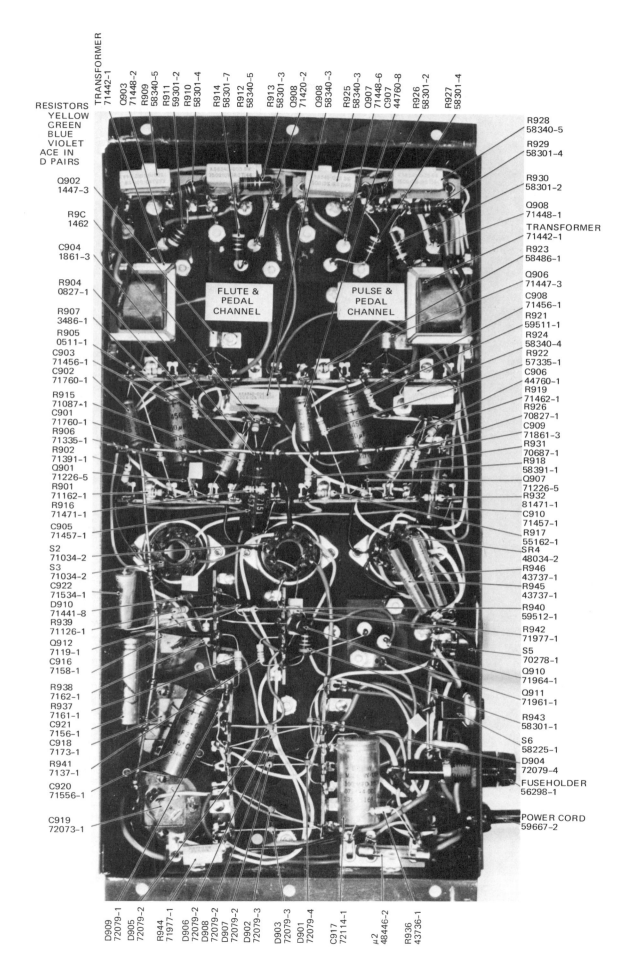

Fig. 10-12 Amplifier power supply

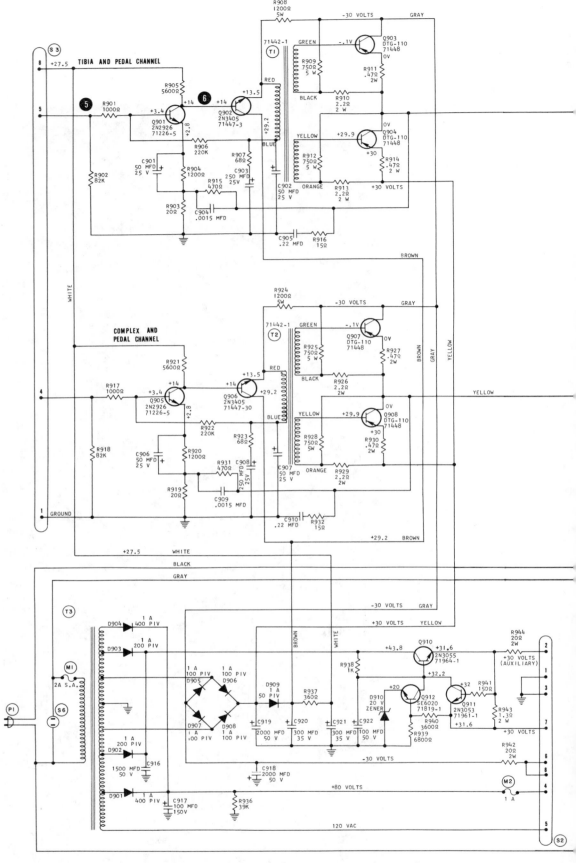

Fig. 10-13 Schematic of figures 10-11 and 10-12

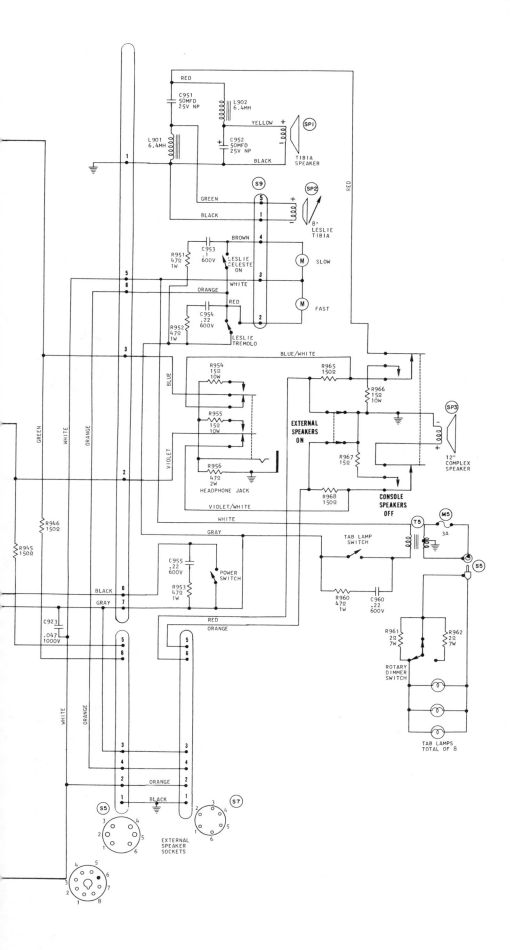

PRINTED CIRCUIT BOARD

Figure 10-14 shows a photograph of a *printed circuit board* with all the components soldered in place. Figure 10-15 shows the *same* circuit board as an X-ray to illustrate how the circuit is connected and where each component is located. Study these two illustrations.

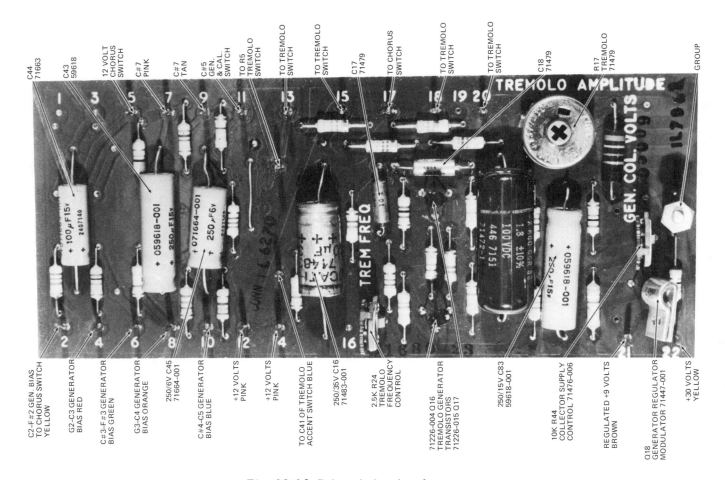

Fig. 10-14 Printed circuit, photo

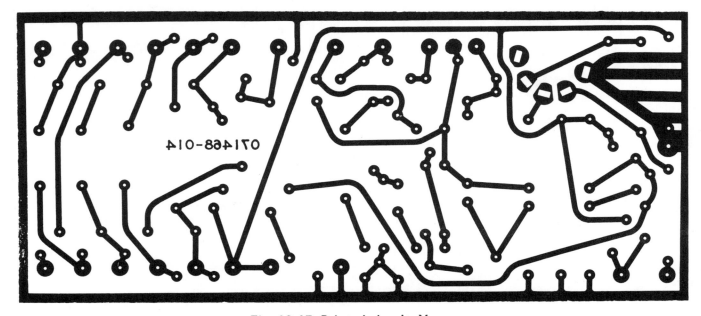

Fig. 10-15 Printed circuit, X-ray

PRINTED CIRCUIT DRAWING

Rules for designing a circuit:

1. Keep conductors as short as possible. Long conductors should be placed near the outside edge of the board.
2. Do not cross conductors over each other. Use a jumper wire.
3. Use correct conductor thickness. For 1.5 amps, make the conductor 0.015 inch wide; for 2.5 amps, make it 0.030 inch wide; for 3.5 amps, make it 0.060 inch wide.
4. Watch the spacing between conductors. For 0-150 volts, use 0.030-inch space; 150-300 volts, use 0.060-inch space; 300-500 volts, use 0.125-inch space.
5. Pads must be large enough to have 0.030 inch of conductor material around the hole in order to provide a good solder base around it. Standard pad size is 0.125 inch.
6. All conductors must join pads smoothly by adding a fillet or round to avoid sharp corners, figure 10-16.
7. Tees and elbows also must have fillets or rounds to avoid sharp corners.

STANDARD PAD   PAD W/STRAIGHT FILLET   DOUBLE PAD

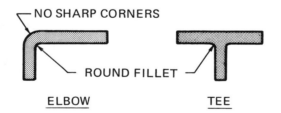

ELBOW   TEE

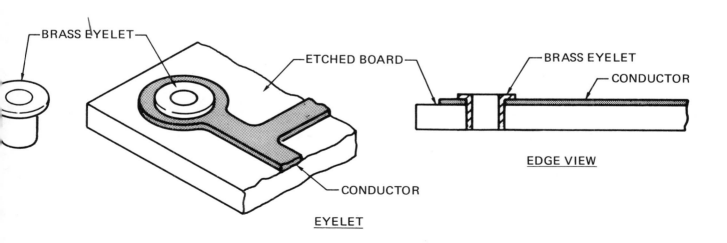

EYELET   EDGE VIEW

Fig. 10-16 Standard pads and conductors

PRINTED CIRCUIT BOARD

Start with an insulating material with copper foil attached, figure 10-17.

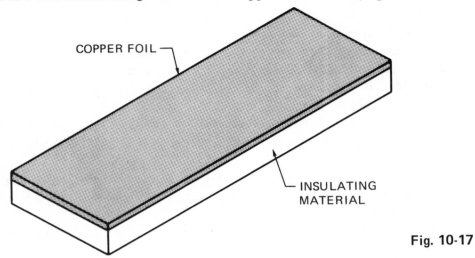

COPPER FOIL

INSULATING MATERIAL

Fig. 10-17

Step 1. Paint the desired pattern on the copper foil. This is often a photograph or silkscreen process using a resist material to protect the copper foil where the resist is, figure 10-18.

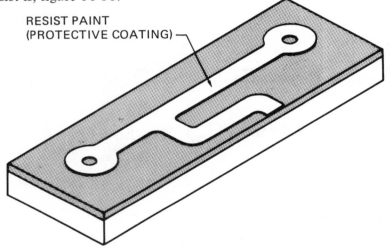

RESIST PAINT (PROTECTIVE COATING)

Fig. 10-18

Step 2. Place the printed circuit (PC) board, with the resist pattern in place, in an acid bath. The acid will etch or eat away the copper foil every place except where the resist pattern is located.

Step 3. Remove the resist paint. There is now a copper circuit where the resist pattern use to be, figure 10-19.

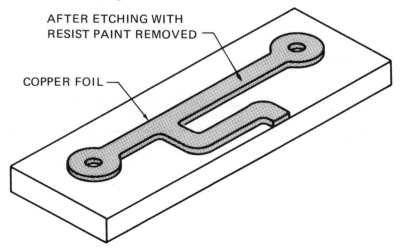

AFTER ETCHING WITH RESIST PAINT REMOVED

COPPER FOIL

Fig. 10-19

This process is like the whiteprint process used to make copies of drawings. The copper foil is like the yellow dye on the copy paper. The drawing's lines represent the resist paint. The light acts like the acid, etching the yellow away. The ammonia developing process is like removing the resist paint.

Schematic

Study the schematic, figure 10-20. To make a printed circuit board, first draw a grid and sketch all components in place (full size). Then, using correct size pads and conductors, make all point-to-point connections. Do not cross any conductors.

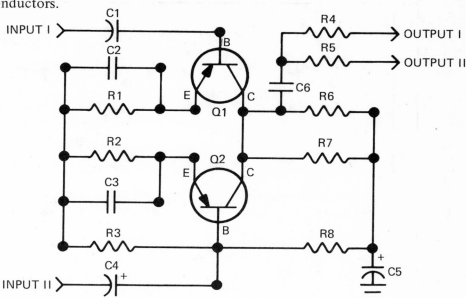

Fig. 10-20 Schematic

Figure 10-21 shows the printed circuit of figure 10-20. Note the actual pads and conductors are on the bottom of the board, the components are on the top. Both top and bottom are illustrated.

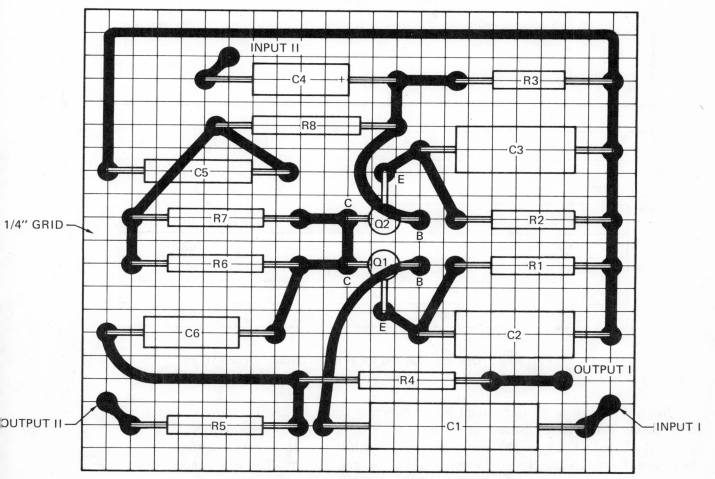

Fig. 10-21 Printed circuit of figure 10-20

DRILLING DRAWING

Carefully study the *drilling drawing* for the printed circuit layout. The holes are dimensioned from the "X" and "Y" axis, numbered left to right and top to bottom, and sizes are indicated, figure 10-22.

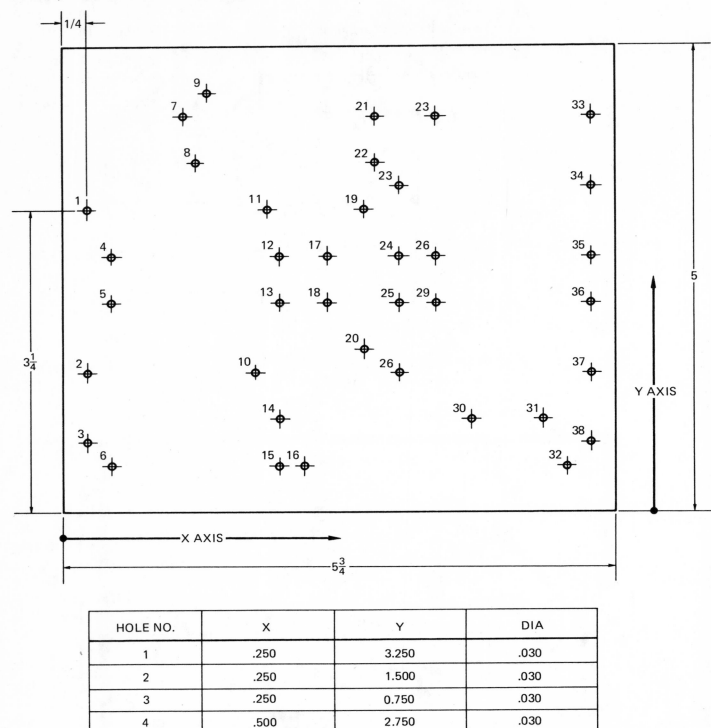

HOLE NO.	X	Y	DIA
1	.250	3.250	.030
2	.250	1.500	.030
3	.250	0.750	.030
4	.500	2.750	.030

Fig. 10-22 Drilling diagram for printed circuit, figure 10-20

Study the finished printed circuit board shown in figure 10-23. This is how the assembly would look. Note that the pads and conductors are on the bottom side of the board. Compare this pictorial drawing with the original schematic of the circuit, figure 10-20. Notice the "pins" for inputs and outputs. This is where wire connections are attached from input and to output.

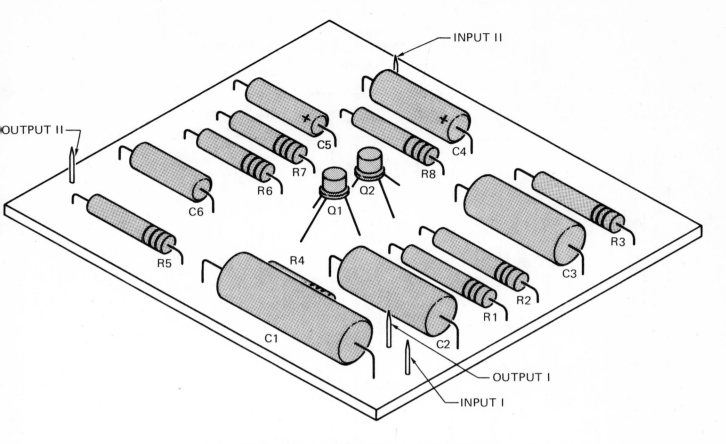

Fig. 10-23 Finished printed circuit, fig. 10-20

Practice Exercise 10-7

Using the schematic, draw a printed circuit board on the grid provided. Check point-to-point contact. The sizes are: R1, R2, R3, R4 (1/4″ × 1 1/4″); C2 (1/4″ × 1″); C1, C3 (1/2″ × 1 1/2″); D1 (1/16″ × 1 1/4″); Q1 (5/16″ diameter); and the four input/output pins (1/16″ square). Draw both the top of the printed circuit board, with all components in place, and the bottom showing all standard 1/4-inch pads, 1/8-inch conductors, and .030-inch diameter holes. Draw a drilling drawing, using "X" and "Y" axis to dimension from, with all holes having a .030-inch diameter. Compare your work to the answer on page 289.

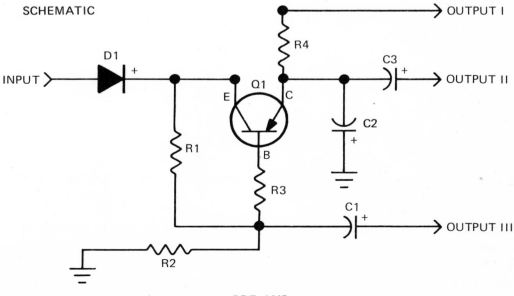

SCHEMATIC

PRE-AMP

GRID

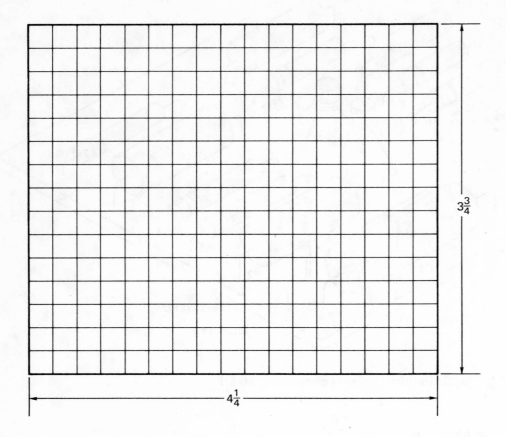

ANSWERS TO PRACTICE EXERCISES

Exercise 10-1

Compare your work with the finished drawing. Notice the dash lines from the antenna, through the amplifier tube VI, and to terminal 5. Do not draw this on the drawing. Notice, also, the power is from terminal 2 to heat the tube's filament. Check all connections. Remember, "V" is the code letter for tubes, "L" for coils, "C" for capacitors, "R" for resistors, etc. Label all components from left to right, as illustrated.

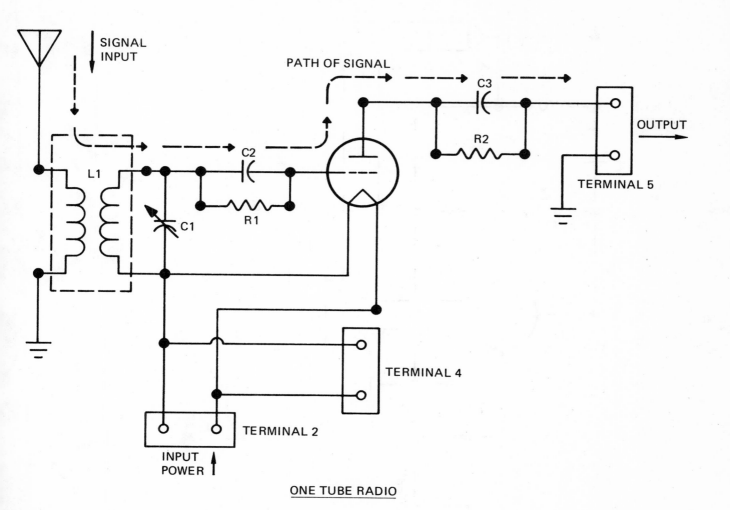

ONE TUBE RADIO

Exercise 10-2

Trace the path of the input signal to the speaker. Check line work and point-to-point connections.

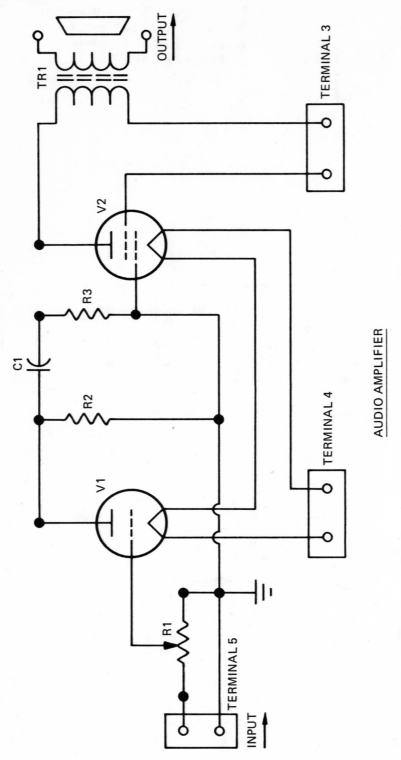

Exercise 10-3

Exercise 10-3 combines all three subassemblies into one unit. Some components have been rearranged to avoid crossovers and to make a neater, well-spaced drawing. Be sure point-to-point contact is exactly the same. Notice how all components are numbered from left to right and how the drawing is read from left to right. Try to follow the audio signal from the antenna, through the amplifier, to the speakers.

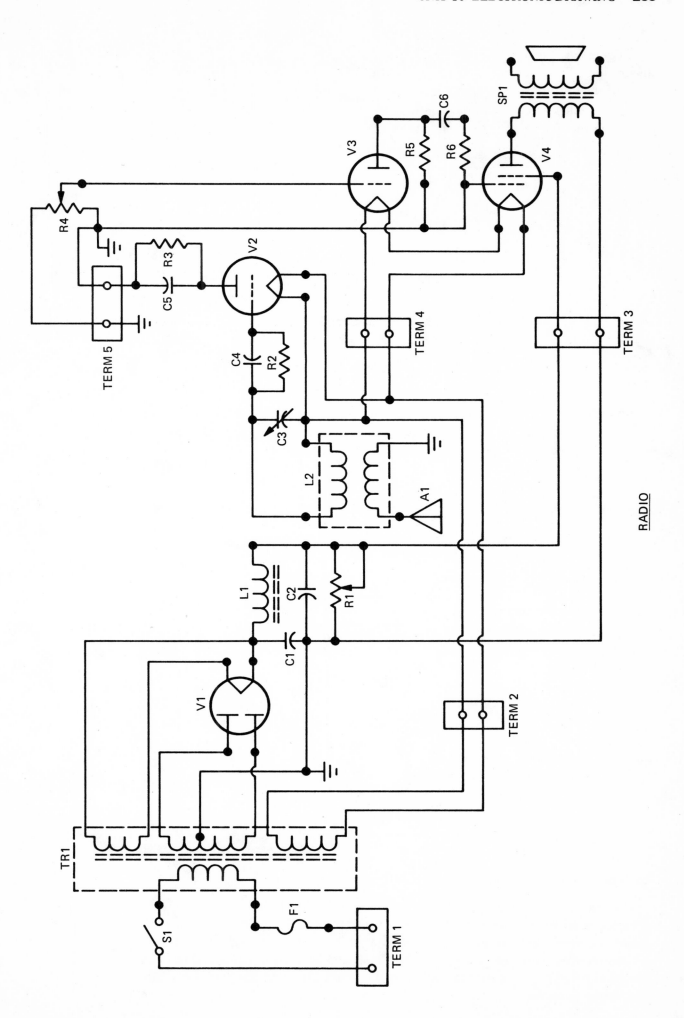

RADIO

Exercise 10-4

Using a 1/2-inch (12) square grid, the practice exercise should look like the illustration. To be correct, the drawing should be reversed with the input to the left and the output on the right. Note point-to-point contacts, line weight, correct symbols, connection points, and neatness. If the input is not on the left and the output on the right, redraw the schematic.

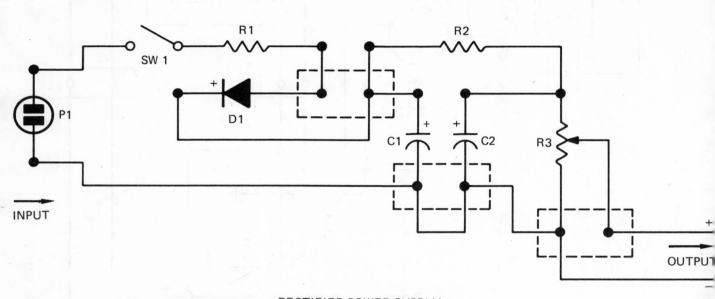

RECTIFIER POWER SUPPLY

Exercise 10-5

Check line work, neatness, point-to-point contacts, and symbols. Input should be on the left and output on the right. Components should be numbered left to right. Try to follow the signal from the antenna through T1 and T2 to the headphones.

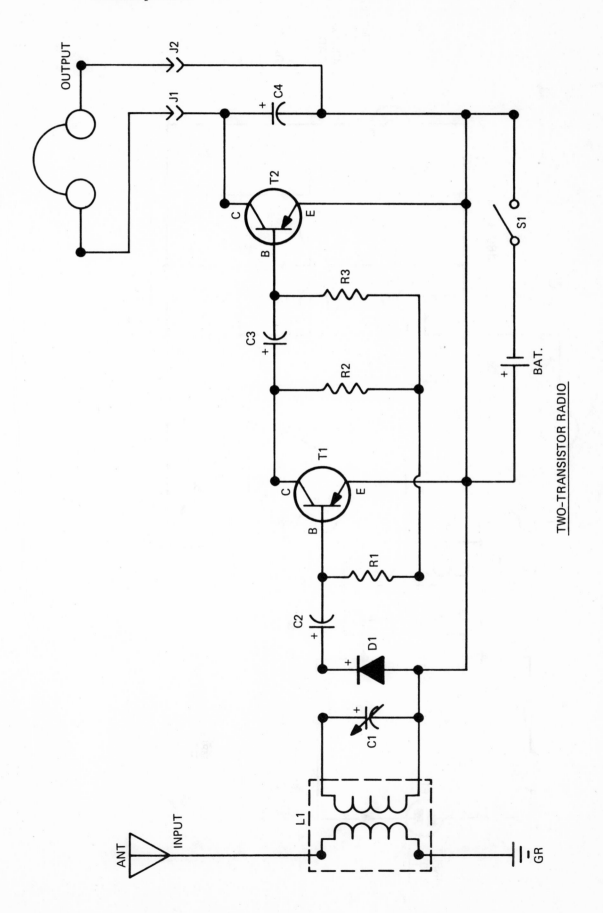

TWO-TRANSISTOR RADIO

Exercise 10-6

Check line work and point-to-point contact. Is the voltmeter across the full load as illustrated?

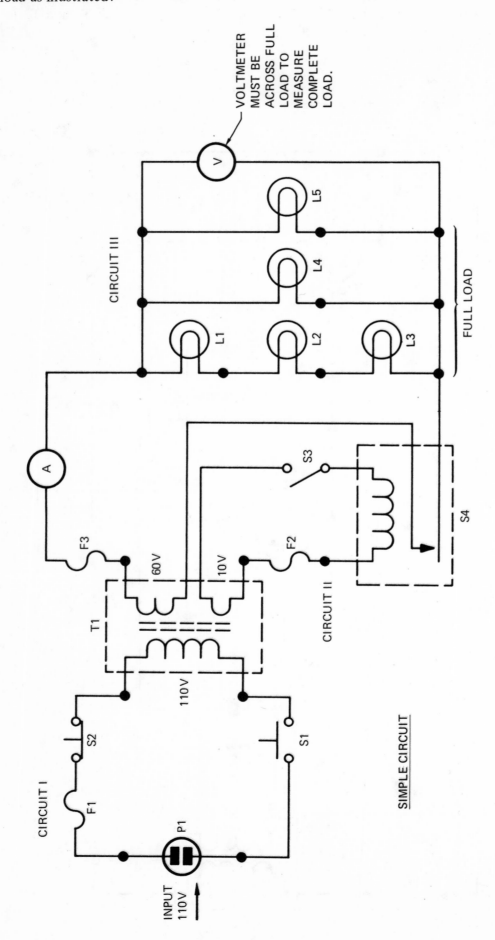

Exercise 10-7

Your drawing may not be exactly the same, but all point-to-point contacts must agree with the schematic drawing. Note neatness, line work, and overall appearance.

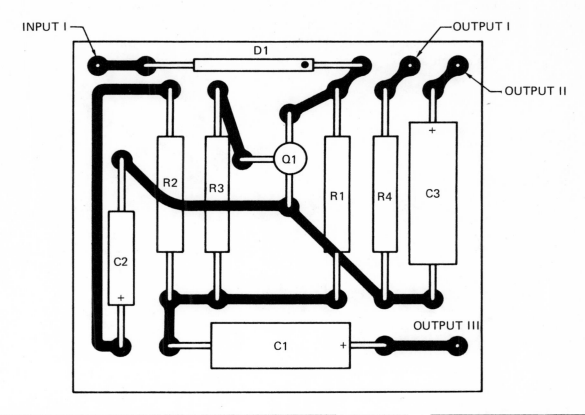

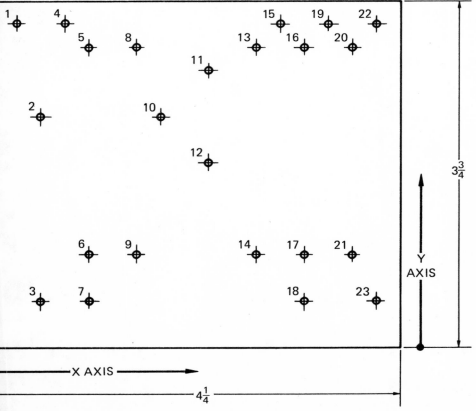

HOLE NO.	X	Y
1	.250	3.50
2	.500	2.50
3	.500	.50
4	.750	3.50
5	1.000	3.25
6	1.000	1.00
7	1.000	.50
8	1.500	3.25
9	1.500	1.00
10	1.750	2.50
11	2.250	3.00
12	2.250	2.00
13	2.750	3.25
14	2.750	1.00
15	3.000	3.50
16	3.250	3.25
17	3.250	1.00
18	3.250	.50
19	3.500	3.50
20	3.750	3.25
21	3.750	1.00
22	4.000	3.50
23	4.000	.50

(ALL HOLES .030 DIA)

UNIT REVIEW

90-minute time limit

A drafter usually gets a rough sketch from the electrical designer. This rough sketch must be redrawn by the electronic drafter using all standard drafting methods. Redraw the rough sketch given on a 3/8-inch (10) grid. Do not forget components may be moved, if necessary, to eliminate extra crossovers. Use correct line weight and standard symbols. Be consistent with all components and make them exactly alike. Label all components from left to right. Be sure to double check point-to-point contacts.

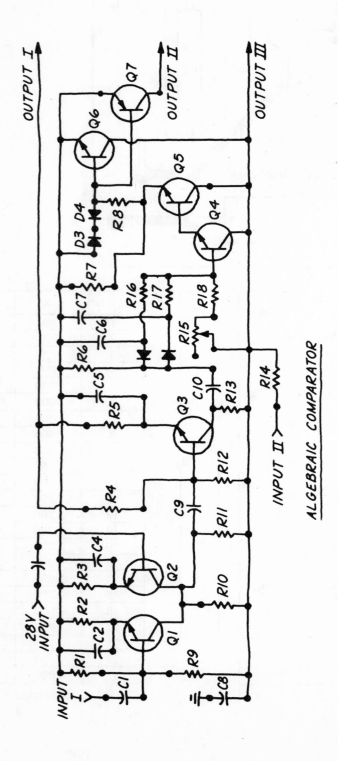

ALGEBRAIC COMPARATOR

Before proceeding to the next unit:

_____ Instructor's approval

_____ Progress plotted

UNIT 11

THE ENGINEERING DEPARTMENT

OBJECTIVE

The student will study the basic functions of an engineering department.

PRETEST

15-minute time limit

1. What is an ECO?

2. What is a parts list? What order should the parts be listed in?

3. What is the normal job progression a student follows, starting from detailer?

4. What should a personal technical file contain?

5. If an employee invents something new, will the company pay him or her for the invention?

6. What is an assembly drawing and does it usually have dimensions?

RELATED TERMS

Give a brief definition of each term as progress is made through the unit.

ECR _____

ECO _____

Parts list _____

Checker _____

Employer/employee agreement _____

A.S.M.E. _____

MIL _____

N.E.M.A. _____

DRAFTING DEPARTMENT PRACTICES

All engineering departments operate differently as each varies in size, personnel, function, and scope. New personnel should fully understand the structure of the organization. Engineering organizations must work as efficiently as possible to provide an orderly flow of drawings. To accomplish this, a uniform procedure or standard is followed by all of its members.

An engineering organization provides drawings for new products, improvement of old products, plans for special custom orders, and many other products. This unit explains some of the basic procedures, paper work, and steps necessary to insure the best efficiency. The following items are included:

- Engineering organization
- How to check a drawing
- Parts list (PL)
- Employer-employee agreement
- Typical engineering project flow chart

ORGANIZATION

Figure 11-1 illustrates a small engineering organization employing 26 people, divided into three organizational elements: detailers, clerks, and typists. In larger organizations comprising one or more departments, depending on the work load, drafters are moved from department to department. The higher the position, the more responsibility and the higher the pay.

CHECKING

Though the drafter is responsible for the accuracy of his work, some companies employ a *checker* to double check all drawings. Some of the things a checker looks for include:

- How is the drawing's general appearance? (legibility, neatness, etc.)
- Does it follow all drawing and company standards?
- Are dimensions and instructions clear and understandable?
- Is the drawing easy to understand?
- Are all dimensions included? A machinist must not have to calculate to find a size or location, assume anything, or have any question whatsoever as to what is required.
- Are there unnecessary dimensions?
- Is the drawing prepared so the part may be manufactured the most economical way?
- Will it assemble with mating parts?
- Have all limits, tolerances, and allowances been properly analyzed for all moving parts?
- Have undesirable accumulations of tolerances been adequately analyzed?
- Are all notes added?
- Are finish texture symbols added?
- Is the material and treatment of each part adequate for the design?
- Is the title block complete? Does it include the title, number of the part, drafter's name, and any other required information?

Because of the high cost of errors, it is important that the drafter check and double check work before releasing the drawing. An engineering drawing must be 100 percent correct.

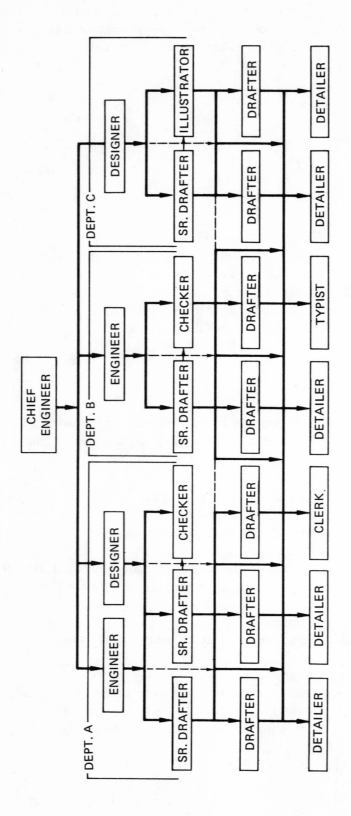

Fig. 11-1 Organization chart

CHANGES

If anyone has a suggestion to improve a part within the product or correct an error on a drawing, it must be brought up at an *engineering change request* (ECR) meeting, figure 11-2. If such suggestions are agreed to by all concerned, an *engineering change order* (ECO) is issued.

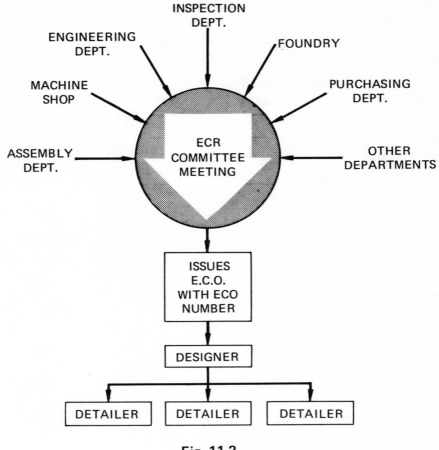

Fig. 11-2

After an engineering change order has been issued, the engineering department assigns a drafter or detailer to make the change(s). After a drawing is issued it cannot be changed by anyone unless it goes through an engineering change request (ECR) and, if approved, an engineering change order (ECO).

In certain cases, administrative changes are appropriate without an engineering change order. Such changes are misspelled words or incorrect references, to improve clarity, etc. These are non-engineering changes. The alternative is to accumulate such changes against the drawing for revision when a more important change is introduced.

The paper work on an engineering change order must include the engineering change request number, why the change was made, who requested the change, what the dimensions were *before* the change, the impact of the change on parts on hand, parts in manufacture, etc.

Change Procedure

After an engineering change order has been issued, the drafter or detailer makes the change.

If the change is extensive: The drawing is completely redrawn and given the same drawing number. The obsolete original is not destroyed but marked "obsolete" and filed.

If the change is fairly simple: The original is carefully erased (so as not to damage the original) where the change is to be made. The change is then carefully added.

If the change is very minor: In the event the change is very minor, such as a small dimension change, the dimension alone is changed and a *wavy line* may be drawn under that dimension indicating it is "out-of-scale." Time is money in an engineering department so all changes must be made in the best and most efficient way. Note that a *change number* is usually placed next to each change and recorded in the *change revision block* on the drawing.

Various companies have different standards or rules governing changes, but the revision block is usually located in a corner of the drawing. It is important that the following is recorded in the revision block.

- Change letter or number (noted beside each change on the drawing also)
- Date of change
- ECR number
- Briefly what was changed
- Name of the person making the change

All changes go through a definite procedure so all departments are notified of a change. Even if the drafter who drew the original finds the error himself after the drawings have been issued, the company's standard change procedure must be followed.

NUMBERING SYSTEMS

All companies have a system of identifying and recording drawings. There is no standard system used by all companies. The most common kind is a letter (A, B, C, D) denoting sheet size, followed by the actual drawing number issued and recorded by a record clerk.

Figure 11-3 shows an example of a detail drawing with four changes (A, B, C, D). Each change was noted by its letter inside a balloon and listed below in the revision record block. The ECO number is also included.

PARTS LIST (PL)

A *parts list* (PL) is a list of all the parts required to assemble a product. It contains the detail number that corresponds to the one on the assembly drawing, the abbreviated drawing title, the plan number, the material, and the quantity required for the assembly. It is the responsibility of the drafter to make up the parts list.

To lay out a parts list:

1. Start with the assembly drawing, first line.
2. List all subassemblies, detail drawings, and purchase parts in the order they are normally assembled, or as you would assemble the unit.
3. Directly after calling out a subassembly, include all detail drawings and purchased parts (indented) that are used to make up that subassembly.
4. Any miscellaneous part (detail drawings, purchased parts) is added at the end of the list.
5. Leave spaces between various subassemblies and parts.

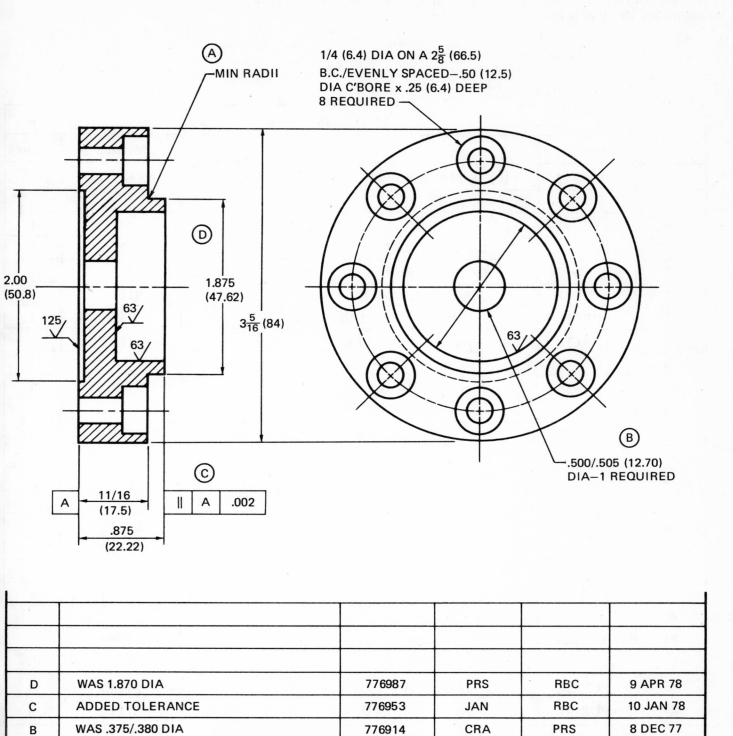

1/4 (6.4) DIA ON A $2\frac{5}{8}$ (66.5)
B.C./EVENLY SPACED—.50 (12.5)
DIA C'BORE x .25 (6.4) DEEP
8 REQUIRED

Ⓐ
—MIN RADII

Ⓓ

2.00
(50.8)

125

63

63

1.875
(47.62)

$3\frac{5}{16}$ (84)

63

Ⓒ

| A | 11/16 (17.5) |

‖ | A | .002

.875
(22.22)

Ⓑ

.500/.505 (12.70)
DIA—1 REQUIRED

D	WAS 1.870 DIA	776987	PRS	RBC	9 APR 78
C	ADDED TOLERANCE	776953	JAN	RBC	10 JAN 78
B	WAS .375/.380 DIA	776914	CRA	PRS	8 DEC 77
A	ADDED NOTE	776872	JAN	RBC	15 NOV 77
REV.	CHANGE	E.C.O. NO.	BY	CHECKED	DATE
REVISION RECORD					

Fig. 11-3 Recording a change

Parts lists vary from company to company, but most look like the sample in figure 11-4. Carefully study the complete list. Note which parts are sub-assemblies, detail drawings, and purchased parts.

PARTS LIST				
No.	Plan No.	Description	Material	Quan.
1	D-77942	VICE ASSEMBLY – MACHINE	AS NOTED	1
2				
3	C-77947	BASE – VICE	AS NOTED	1
4	B-77952	BASE – LOWER	C.I.	1
5	A-77951	BASE – UPPER	C.I.	1
6	A-77946	SPACER – BASE	STEEL	1
7	PURCH.	BOLT – ½ – 13 UNC – 2″ LG.	STEEL	1
8	PURCH.	NUT – ½ – 13 UNC	STEEL	1
9				
10	C-77955	JAW – SLIDING	STEEL	1
11				
12	A-77954	SCREW – VICE	STEEL	1
13	A-77953	ROD – HANDLE	STEEL	1
14	A-77956	BALL – HANDLE	STEEL	2
15				
16	A-77961	PLATE – JAN	STEEL	2
17	PURCH.	SCREW – ¼ – 20 UNC	STEEL	4
18				
19	A-77962	COLLAR	STEEL	1
20				
21	A-77841	KEY – SPECIAL VICE	STEEL	2
22				
23	PURCH.	BOLT – ½ – 13 UNC – 4″ LG.	STEEL	4
24	PURCH.	NUT – ½ – 13 UNC	STEEL	4

Company Name Company Address	Model No. 999	Parts Lister J.A.N.	Date 5 MAR. 79
Title VICE ASSEMBLY – MACHINE	Page _1_ of _1_ Pages		DWG No. D-77942

Fig. 11-4 Parts list

Practice Exercise 11-1

Using the sample parts list in figure 11-4, answer the following questions. Compare your work to the answers on page 305.

1. How many parts are purchased? Which line numbers are they?

2. How many subassemblies are there? Which line numbers are they?

3. What is the general order of the parts?

4. How many detail parts are there? Which line numbers are they?

5. What is the model number? What is the title?

6. Which part numbers make up the screw-vice? How many parts are there to that subassembly?

7. What is the plan number for line number 17?

8. What material is used in the C-77947 subassembly?

9. What size paper is the assembly drawing done on?

EMPLOYER-EMPLOYEE AGREEMENT

Usually companies that work at designing or creating new devices or processes will have all employees sign an *employer/employee agreement.* Simply stated:

1. You must promise not to reveal any company secrets.
2. If you invent something or discover some new process or method, it belongs to the company.
3. You will not attempt to infringe on any company ideas, processes, or inventions.

This agreement is usually in effect while you are employed and extended for a given time after you leave the company, often six months to two years.

As a rule, you do not get extra compensation for a new design or process, but most companies recognize your efforts and in time adjust your salary or position. In many instances a bonus payment is made in the form of bonds, company stocks, or other forms of recognition.

VARIOUS STANDARDS USED

Companies use various engineering rules or standards. A few such standards include:

- A.N.S.I. – American National Standards Institute
- S.A.E. – Society of Automotive Engineers
- MIL. – Military Standards
- A.S.M.E. – American Society of Mechanical Engineers
- N.E.M.A. – National Electrical Manufacturers Association

Each company also has its own rules or standards to follow.

PERSONAL TECHNICAL FILE

It is important to locate technical information quickly. Usually a company manufacturing a certain product or performing a particular kind of engineering will use certain technical information. A conscientious drafter should develop and update a *personal technical file* containing:

- All company products associated with assignments
- Notes, copies, or clippings from various technical magazines, literature, etc. associated with the company product
- Information on standard materials commonly used in assignments
- Miscellaneous information that would make your job more efficient
- Records of various supervisors, pay levels, and dates of special assignments.
- Various pages of units in this book that were helpful. This material should be neatly organized in a loose leaf binder.

PROJECT FLOW CHART

The process of developing a new product from its established need and conception to its final production consumes many hours and involves highly trained personnel. The flow chart in figure 11-5 gives an example of the process involved.

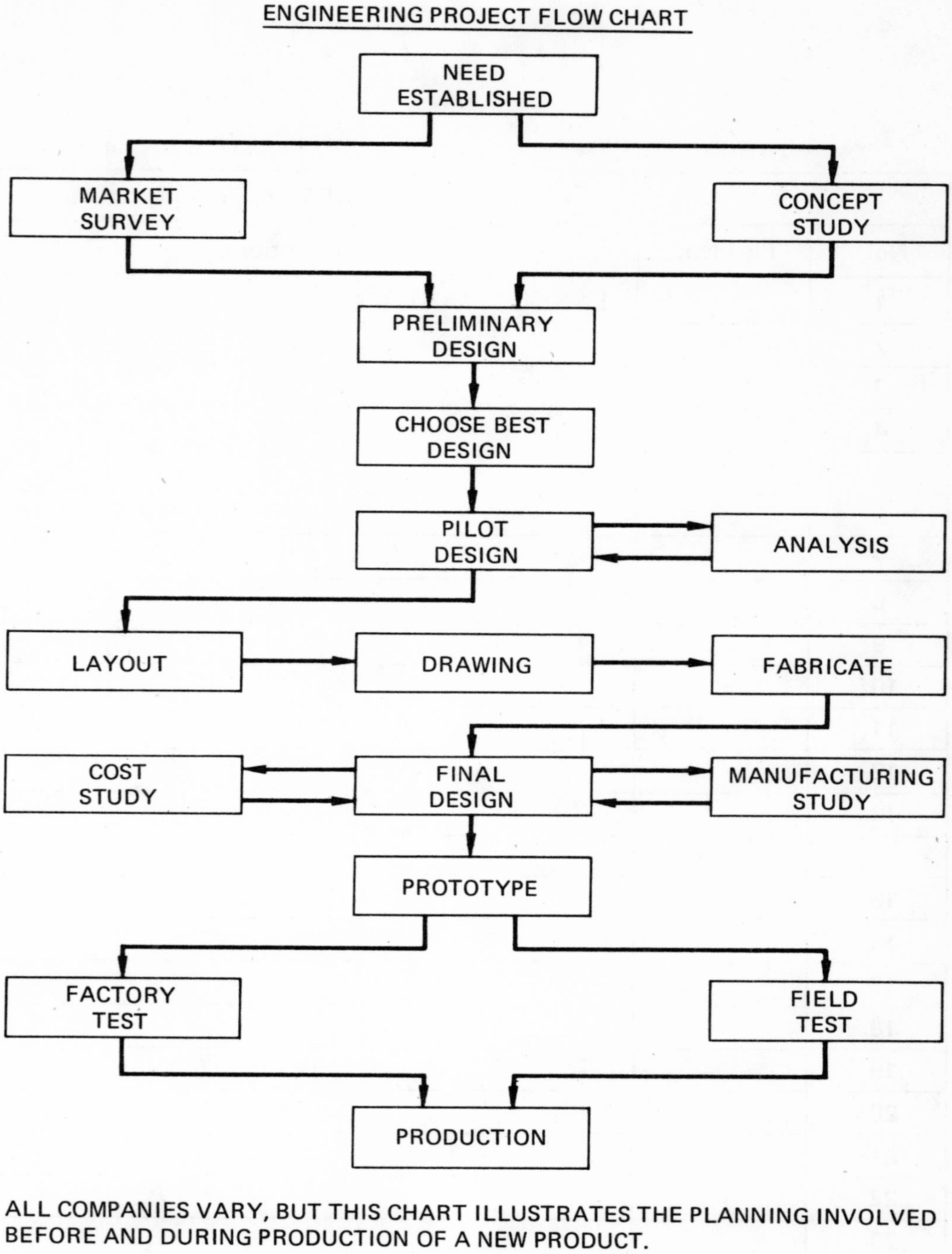

ENGINEERING PROJECT FLOW CHART

```
                    ┌──────────────┐
                    │    NEED      │
                    │ ESTABLISHED  │
                    └──────────────┘
           ┌───────────────┴───────────────┐
    ┌──────────────┐                ┌──────────────┐
    │   MARKET     │                │   CONCEPT    │
    │   SURVEY     │                │    STUDY     │
    └──────────────┘                └──────────────┘
           └───────────────┬───────────────┘
                    ┌──────────────┐
                    │ PRELIMINARY  │
                    │   DESIGN     │
                    └──────────────┘
                           │
                    ┌──────────────┐
                    │ CHOOSE BEST  │
                    │   DESIGN     │
                    └──────────────┘
                           │
                    ┌──────────────┐        ┌──────────────┐
                    │    PILOT     │───────▶│              │
                    │   DESIGN     │◀───────│   ANALYSIS   │
                    └──────────────┘        └──────────────┘
        ┌──────────────────┘
 ┌──────────────┐    ┌──────────────┐    ┌──────────────┐
 │    LAYOUT    │───▶│   DRAWING    │───▶│  FABRICATE   │
 └──────────────┘    └──────────────┘    └──────────────┘
                            ┌───────────────┘
 ┌──────────────┐    ┌──────────────┐    ┌──────────────┐
 │    COST      │◀───│    FINAL     │───▶│ MANUFACTURING│
 │    STUDY     │───▶│   DESIGN     │◀───│    STUDY     │
 └──────────────┘    └──────────────┘    └──────────────┘
                    ┌──────────────┐
                    │  PROTOTYPE   │
                    └──────────────┘
           ┌───────────────┴───────────────┐
    ┌──────────────┐                ┌──────────────┐
    │   FACTORY    │                │    FIELD     │
    │    TEST      │                │    TEST      │
    └──────────────┘                └──────────────┘
           └───────────────┬───────────────┘
                    ┌──────────────┐
                    │  PRODUCTION  │
                    └──────────────┘
```

ALL COMPANIES VARY, BUT THIS CHART ILLUSTRATES THE PLANNING INVOLVED BEFORE AND DURING PRODUCTION OF A NEW PRODUCT.

Fig. 11-5 Project flow chart

Practice Exercise 11-2

Complete the parts list using the model airplane engine assembly, subassembly, and detail drawings in Unit 8, Exercise 8-2. Compare your work to the answers on page 305.

PARTS LIST					
No.	Plan No.	Description		Material	Quan.
1	–	ENGINE – MODEL AIRPLANE		AS NOTED	1
2					
3					
4					
5					
6					
7					
8					
9					
10					
11					
12					
13					
14					
15					
16					
17					
18					
19					
20					
21					
22					
23					
24					
Company Name Company Address		Model No.	Parts Lister		Date
Title			Page ___ of ___ Pages		DWG. No.

ANSWERS TO PRACTICE EXERCISES

Carefully compare your work to the answer for each exercise. Refer any questions to the instructor.

Exercise 11-1

1. Five purchased parts; line numbers 7, 8, 17, 23, 24.
2. Three subassemblies; line numbers 3, 12, 16.
3. Place the parts in the order recommended or suggested for assembly.
4. Eight detail parts; line numbers 4, 5, 6, 10, 13, 14, 19, 21.
5. 999 model number; vice assembly, machine.
6. (One) A-77953 and (two) A-77956; three parts total.
7. No plan number, purchased parts are listed only as "purch."
8. C.I. (base-lower) C.I. (base-upper) steel (spacer-base) and steel (bolt/nut) "as noted" indicates each part in the assembly is noted in its own call-off.
9. "D" size or 22″ X 34″ (560 mm X 865 mm).

Exercise 11-2

The list does not have to exactly be in this order but must contain all drawings.

PARTS LIST				
No.	Plan No.	Description	Material	Quan.
1	—	ENGINE – MODEL AIRPLANE	AS NOTED	1
2	3	CRANKCASE SUB-ASSEMBLY	AS NOTED	1
3	1	CRANKCASE	ALUM	1
4	2	BUSHING – CRANKCASE	LEAD/BRON	1
5	4	CRANKSHAFT	STEEL	1
6	5	ROD – CONNECTING	ALUM	1
7	8	PIN – DRIVEWASHER	STEEL	1
8	9	PIN – PISTON	STEEL	1
9	10	SPACER – ROD	BRASS	2
10	11	BACKPLATE	ALUM	1
11	PURCH.	SCREW – FILL HD. MACH.	—	—
12	—	6–32 UNC – 2A x 7/16 LG.	STEEL	4
13	13	SLEEVE – CYLINDER	IRON	1
14	12	FINS – CYLINDER	ALUM	1
15	14	PISTON	IRON	1
16	15	HEAD – CYLINDER	ALUM	1
17	PURCH.	SCREW – FILL HD. MACH.	—	—
18	—	6–32 UNC – 2A x 1 3/16 LG.	STEEL	3
19	PURCH.	SCREW – FILL HD. MACH.	—	—
20	—	6–32 UNC –2A x 5/16 LG.	STEEL	3
21	6	DRIVEWASHER	STEEL	1
22	7	SPINNER	ALUM	1
23				
24				

Company Name Company Address	Model No. .60	Parts Lister P.R.S.	Date 6 MAR 79
Title ENGINE – MODEL AIRPLANE	Page _1_ of _1_ Pages		DWG. No.

UNIT REVIEW

30-minute time limit

1. Who makes up an ECR committee?

2. Why does a company purchase standard items?

3. What number is usually used on a parts list?

4. What does the letter "B" in front of a plan number indicate?

5. What does a checker do? List some of the duties?

6. Briefly explain how a change is made on a drawing.

7. What does a change revision block contain? Where is it located?

8. What are three engineering standards that a company may follow?

9. What does a wavy line under a dimension on a drawing indicate?

10. What is an employer-employee agreement?

_____ Instructor's approval

_____ Progress plotted

ACKNOWLEDGMENTS

The author wishes to thank the following for reviewing the manuscript and providing critical input:

Robert Franciose, Chairman
ANSI Drafting Standards (Y14) Committee

Delmar Staff

Industrial Education Editor — Mark W. Huth
Associate Editor — Kathleen E. Beiswenger
Technical Editor — Harry A. Sturges

Illustrations

Conn Organ Corporation 10-2, 10-3, 10-4, 10-5, 10-11, 10-12, 10-13, 10-14, 10-15
Gar Electroforming, Division of Mite Company 7-26
Heath Company 9-32
Lockheed Missiles & Space Company title page
L.S. Starrett Company 3-17, 4-1, 4-2, 4-3, 4-4, 4-5, 4-6, 4-7, 4-8, 4-9, 4-10, 4-11, 4-13
Triplety Corporation 9-33, 9-34, 9-35

Classroom Testing

The instructional material in this text was classroom tested in the vocational drafting department at St. Johnsbury Academy, St. Johnsbury, Vermont.

APPENDIX A

Fraction	Decimal Equivalent		Fraction	Decimal Equivalent	
	Customary (in.)	Metric (mm)		Customary (in.)	Metric (mm)
1/64——.015625	.015625	0.3969	33/64——.515625	.515625	13.0969
1/32——.03125	.03125	0.7938	17/32——.53125	.53125	13.4938
3/64——.046875	.046875	1.1906	35/64——.546875	.546875	13.8906
1/16——.0625	.0625	1.5875	9/16——.5625	.5625	14.2875
5/64——.078125	.078125	1.9844	37/64——.578125	.578125	14.6844
3/32——.09375	.09375	2.3813	19/32——.59375	.59375	15.0813
7/64——.109375	.109375	2.7781	39/64——.609375	.609375	15.4781
1/8——.1250	.1250	3.1750	5/8——.6250	.6250	15.8750
9/64——.140625	.140625	3.5719	41/64——.640625	.640625	16.2719
5/32——.15625	.15625	3.9688	21/32——.65625	.65625	16.6688
11/64——.171875	.171875	4.3656	43/64——.671875	.671875	17.0656
3/16——.1875	.1875	4.7625	11/16——.6875	.6875	17.4625
13/64——.203125	.203125	5.1594	45/64——.703125	.703125	17.8594
7/32——.21875	.21875	5.5563	23/32——.71875	.71875	18.2563
15/64——.234375	.234375	5.9531	47/64——.734375	.734375	18.6531
1/4——.250	.250	6.3500	3/4——.750	.750	19.0500
17/64——.265625	.265625	6.7469	49/64——.765625	.765625	19.4469
9/32——.28125	.28125	7.1438	25/32——.78125	.78125	19.8438
19/64——.296875	.296875	7.5406	51/64——.796875	.796875	20.2406
5/16——.3125	.3125	7.9375	13/16——.8125	.8125	20.6375
21/64——.328125	.328125	8.3384	53/64——.828125	.828125	21.0344
11/32——.34375	.34375	8.7313	27/32——.84375	.84375	21.4313
23/64——.359375	.359375	9.1281	55/64——.859375	.859375	21.8281
3/8——.3750	.3750	9.5250	7/8——.8750	.8750	22.2250
25/64——.390625	.390625	9.9219	57/64——.890625	.890625	22.6219
13/32——.40625	.40625	10.3188	29/32——.90625	.90625	23.0188
27/64——.421875	.421875	10.7156	59/64——.921875	.921875	23.4156
7/16——.4375	.4375	11.1125	15/16——.9375	.9375	23.8125
29/64——.453125	.453125	11.5094	61/64——.953125	.953125	24.2094
15/32——.46875	.46875	11.9063	31/32——.96875	.96875	24.6063
31/64——.484375	.484375	12.3031	63/64——.984375	.984375	25.0031
1/2——.500	.500	12.7000	1——1.000	1.000	25.4000

INCH/METRIC EQUIVALENTS

APPENDIX B

DIMENSION AND SIZE CHART FOR THREADS

Nominal Size		Diameter (Major)		Diameter (Minor)		Tap Drill (For 75% Th'd.)			Threads Per Inch		Pitch (MM)		T.P.I. (Approx.)	
Inch	M.M.	Inch	M.M.	Inch	M.M.	Drill	Inch	M.M.	UNC	UNF	Coarse	Fine	Coarse	Fine
—	M1.4	.055	1.397	—	—	—	—	—	—	—	.3	.2	85	127
0	—	.060	1.524	.0438	1.092	3/64	.0469	1.168	—	80	—	—	—	—
—	M1.6	.063	1.600	—	—	—	—	—	—	—	.35	.2	74	127
1	—	.073	1.854	.0527	1.320	53	.0595	1.499	64	—	—	—	—	—
1	—	.073	1.854	.0550	1.397	53	.0595	1.499	—	72	—	—	—	—
—	M.2	.079	2.006	—	—	—	—	—	—	—	.4	.25	64	101
2	—	.086	2.184	.0628	1.587	50	.0700	1.778	56	—	—	—	—	—
2	—	.086	2.184	.0657	1.651	50	.0700	1.778	—	64	—	—	—	—
—	M2.5	.098	2.489	—	—	—	—	—	—	—	.45	.35	56	74
3	—	.099	2.515	.0719	1.828	47	.0785	1.981	48	—	—	—	—	—
3	—	.099	2.515	.0758	1.905	46	.0810	2.057	—	58	—	—	—	—
4	—	.112	2.845	.0795	2.006	43	.0890	2.261	40	—	—	—	—	—
4	—	.112	2.845	.0849	2.134	42	.0935	2.380	—	48	—	—	—	—
—	M3	.118	2.997	—	—	—	—	—	—	—	.5	.35	51	74
5	—	.125	3.175	.0925	2.336	38	.1015	2.565	40	—	—	—	—	—
5	—	.125	3.175	.9055	2.413	37	.1040	2.641	—	44	—	—	—	—
6	—	.138	3.505	.0975	2.464	36	.1065	2.692	32	—	—	—	—	—
6	—	.138	3.505	.1055	2.667	33	.1130	2.870	—	40	—	—	—	—
—	M4	.157	3.988	—	—	—	—	—	—	—	.7	.35	36	51
8	—	.164	4.166	.1234	3.124	29	.1360	3.454	32	—	—	—	—	—
8	—	.164	4.166	.1279	3.225	29	.1360	3.454	—	36	—	—	—	—
10	—	.190	4.826	.1359	3.429	26	.1470	3.733	24	—	—	—	—	—
10	—	.190	4.826	.1494	3.785	21	.1590	4.038	—	32	—	—	—	—
—	M5	.196	4.978	—	—	—	—	—	—	—	.8	.5	32	51
12	—	.216	5.486	.1619	4.089	16	.1770	4.496	24	—	—	—	—	—
12	—	.216	5.486	.1696	4.293	15	.1800	4.572	—	28	—	—	—	—
—	M6	.236	5.994	—	—	—	—	—	—	—	1.0	.75	25	34
1/4	—	.250	6.350	.1850	4.699	7	.2010	5.105	20	—	—	—	—	—
1/4	—	.250	6.350	.2036	5.156	3	.2130	5.410	—	28	—	—	—	—
5/16	—	.312	7.938	.2403	6.096	F	.2570	6.527	18	—	—	—	—	—
5/16	—	.312	7.938	.2584	6.553	I	.2720	6.908	—	24	—	—	—	—
—	M8	.315	8.001	—	—	—	—	—	—	—	1.25	1.0	20	25
3/8	—	.375	9.525	.2938	7.442	5/16	.3125	7.937	16	—	—	—	—	—
3/8	—	.375	9.525	.3209	8.153	Q	.3320	8.432	—	24	—	—	—	—
—	M10	.393	9.982	—	—	—	—	—	—	—	1.5	1.25	17	20
7/16	—	.437	11.113	.3447	8.738	U	.3680	9.347	14	—	—	—	—	—
7/16	—	.437	11.113	.3726	9.448	25/64	.3906	9.921	—	20	—	—	—	—
—	M12	.471	11.963	—	—	—	—	—	—	—	1.75	1.25	14.5	20
1/2	—	.500	12.700	.4001	10.162	27/64	.4219	10.715	13	—	—	—	—	—
1/2	—	.500	12.700	.4351	11.049	29/64	.4531	11.509	—	20	—	—	—	—
—	M14	.551	13.995	—	—	—	—	—	—	—	2	1.5	12.5	17
9/16	—	.562	14.288	.4542	11.531	31/64	.4844	12.3031	12	—	—	—	—	—
9/16	—	.562	14.288	.4903	12.446	33/64	.5156	13.096	—	18	—	—	—	—
5/8	—	.625	15.875	.5069	12.852	17/32	.5312	13.493	11	—	—	—	—	—
5/8	—	.625	15.875	.5528	14.020	37/64	.5781	14.684	—	18	—	—	—	—
—	M16	.630	16.002	—	—	—	—	—	—	—	2	1.5	12.5	17
—	M18	.709	18.008	—	—	—	—	—	—	—	2.5	1.5	10	17
3/4	—	.750	19.050	.6201	15.748	21/32	.6562	16.668	10	—	—	—	—	—
3/4	—	.750	19.050	.6688	16.967	11/16	.6875	17.462	—	16	—	—	—	—
—	M20	.787	19.990	—	—	—	—	—	—	—	2.5	1.5	10	17
—	M22	.866	21.996	—	—	—	—	—	—	—	2.5	1.5	10	17
7/8	—	.875	22.225	.7307	18.542	49/64	.7656	19.446	9	—	—	—	—	—
7/8	—	.875	22.225	.7822	19.863	13/16	.8125	20.637	—	14	—	—	—	—
—	M24	.945	24.003	—	—	—	—	—	—	—	3	2	8.5	12.5
1	—	1.000	25.400	.8376	21.2598	7/8	.8750	22.225	8	—	—	—	—	—
1	—	1.000	25.400	.8917	22.632	59/64	.9219	23.415	—	12	—	—	—	—
—	M27	1.063	27.000	—	—	—	—	—	—	—	3	2	8.5	12.5

VALUES IN THOUSANDTHS OF AN INCH

Nominal Size Range Inches		Class RC1 Precision Sliding			Class RC2 Sliding Fit			Class RC3 Precision Running			Class RC4 Close Running			Class RC5 Medium Running		
		Hole Tol. GR5	Minimum Clearance	Shaft Tol. GR4	Hole Tol. GR6	Minimum Clearance	Shaft Tol. GR5	Hole Tol. GR7	Minimum Clearance	Shaft Tol. GR6	Hole Tol. GR8	Minimum Clearance	Shaft Tol. GR7	Hole Tol. GR8	Minimum Clearance	Shaft Tol. GR7
Over	To	-0		+0	-0		+0	-0		+0	-0		+0	-0		+0
0	.12	+0.15	0.10	-0.12	+0.25	0.10	-0.15	+0.40	0.30	-0.25	+0.60	0.30	-0.40	+0.60	0.60	-0.40
.12	.24	+0.20	0.15	-0.15	+0.30	0.15	-0.20	+0.50	0.40	-0.30	+0.70	0.40	-0.50	+0.70	0.80	-0.50
.24	.40	+0.25	0.20	-0.15	+0.40	0.20	-0.25	+0.60	0.50	-0.40	+0.90	0.50	-0.60	+0.90	1.00	-0.60
.40	.71	+0.30	0.25	-0.20	+0.40	0.25	-0.30	+0.70	0.60	-0.40	+1.00	0.60	-0.70	+1.00	1.20	-0.70
.71	1.19	+0.40	0.30	-0.25	+0.50	0.30	-0.40	+0.80	0.80	-0.50	+1.20	0.80	-0.80	+1.20	1.60	-0.50
1.19	1.97	+0.40	0.40	-0.30	+0.60	0.40	-0.40	+1.00	1.00	-0.60	+1.60	1.00	-1.00	+1.60	2.00	-1.00
1.97	3.15	+0.50	0.40	-0.30	+0.70	0.40	-0.50	+1.20	1.20	-0.70	+1.80	1.20	-1.20	+1.80	2.50	-1.20
3.15	4.73	+0.60	0.50	-0.40	+0.90	0.50	-0.60	+1.40	1.40	-0.90	+2.20	1.40	-1.40	+2.20	3.00	-1.40
4.73	7.09	+0.70	0.60	-0.50	+1.00	0.60	-0.70	+1.60	1.60	-1.00	+2.50	1.60	-1.60	+2.50	3.50	-1.60
7.09	9.85	+0.80	0.60	-0.60	+1.20	0.60	-0.80	+1.80	2.00	-1.20	+2.80	2.00	-1.80	+2.80	4.50	-1.80
9.85	12.41	+0.90	0.80	-0.60	+1.20	0.80	-0.90	+2.00	2.50	-1.20	+3.00	2.50	-2.00	+3.00	5.00	-2.00
12.41	15.75	+1.00	1.00	-0.70	+1.40	1.00	-1.00	+2.20	3.00	-1.40	+3.50	3.00	-2.20	+3.50	6.00	-2.20

Nominal Size Range Inches		Class RC6 Medium Running			Class RC7 Free Running			Class RC8 Loose Running			Class RC9 Loose Running		
		Hole Tol. GR9	Minimum Clearance	Shaft Tol. GR8	Hole Tol. GR9	Minimum Clearance	Shaft Tol. GR8	Hole Tol. GR10	Minimum Clearance	Shaft Tol. GR9	Hole Tol. GR11	Minimum Clearance	Shaft Tol. GR10
Over	To	-0		+0	-0		+0	-0		+0	-0		+0
0	.12	+1.00	0.60	-0.60	+1.00	1.00	-0.60	+1.60	2.50	-1.00	+2.50	4.00	-1.60
.12	.24	+1.20	0.80	-0.70	+1.20	1.20	-0.70	+1.80	2.80	-1.20	+3.00	4.50	-1.80
.24	.40	+1.40	1.00	-0.90	+1.40	1.60	-0.90	+2.20	3.00	-1.40	+3.50	6.00	-2.20
.40	.71	+1.60	1.20	-1.00	+1.60	2.00	-1.00	+2.80	3.50	-1.60	+4.00	6.00	-2.80
.71	1.19	+2.00	1.60	-1.20	+2.00	2.50	-1.20	+3.50	4.50	-2.00	+5.00	7.00	-3.50
1.19	1.97	+2.50	2.00	-1.60	+2.50	3.00	-1.60	+4.00	5.00	-2.50	+6.00	8.00	-4.00
1.97	3.15	+3.00	2.50	-1.80	+3.00	4.00	-1.80	+4.50	6.00	-3.00	+7.00	9.00	-4.50
3.15	4.73	+3.50	3.00	-2.20	+3.50	5.00	-2.20	+5.00	7.00	-3.50	+9.00	10.00	-5.00
4.73	7.09	+4.00	3.50	-2.50	+4.00	6.00	-2.50	+6.00	8.00	-4.00	+10.00	12.00	-6.00
7.09	9.85	+4.50	4.00	-2.80	+4.50	7.00	-2.80	+7.00	10.00	-4.50	+12.00	15.00	-7.00
9.85	12.41	+5.00	5.00	-3.00	+5.00	8.00	-3.00	+8.00	12.00	-5.00	+12.00	18.00	-8.00
12.41	15.75	+6.00	6.00	-3.50	+6.00	10.00	-3.50	+9.00	14.00	-6.00	+14.00	22.00	-9.00

VALUES IN MILLIMETRES

Nominal Size Range Millimetres		Class RC1 Precision Sliding			Class RC2 Sliding Fit			Class RC3 Precision Running			Class RC4 Close Running			Class RC5 Medium Running		
		Hole Tol. H5	Minimum Clearance	Shaft Tol. g4	Hole Tol. H6	Minimum Clearance	Shaft Tol. g5	Hole Tol. H7	Minimum Clearance	Shaft Tol. f6	Hole Tol. H8	Minimum Clearance	Shaft Tol. f7	Hole Tol. H8	Minimum Clearance	Shaft Tol. e7
Over	To	-0		+0	-0		+0	-0		+0	-0		+0	-0		+0
0	3	+0.004	0.003	-0.003	+0.006	0.003	-0.004	+0.010	0.008	-0.006	+0.015	0.008	-0.010	+0.015	0.015	-0.010
3	6	+0.005	0.004	-0.004	+0.008	0.004	-0.005	+0.013	0.010	-0.008	+0.018	0.010	-0.013	+0.018	0.020	-0.013
6	10	+0.006	0.005	-0.004	+0.010	0.005	-0.006	+0.015	0.013	-0.010	+0.023	0.013	-0.015	+0.018	0.025	-0.015
10	18	+0.008	0.006	-0.005	+0.010	0.006	-0.008	+0.018	0.015	-0.010	+0.025	0.015	-0.018	+0.025	0.030	-0.018
18	30	+0.010	0.008	-0.006	+0.013	0.008	-0.010	+0.020	0.020	-0.013	+0.030	0.020	-0.020	+0.030	0.040	-0.020
30	50	+0.010	0.010	-0.008	+0.015	0.010	-0.010	+0.030	0.030	-0.015	+0.040	0.030	-0.030	+0.040	0.050	-0.030
50	80	+0.013	0.010	-0.008	+0.018	0.010	-0.013	+0.030	0.030	-0.020	+0.050	0.030	-0.030	+0.050	0.060	-0.030
80	120	+0.015	0.013	-0.010	+0.023	0.013	-0.015	+0.040	0.040	-0.020	+0.060	0.040	-0.040	+0.060	0.080	-0.040
120	180	+0.018	0.015	-0.013	+0.025	0.015	-0.018	+0.040	0.040	-0.030	+0.060	0.040	-0.040	+0.060	0.090	-0.040
180	250	+0.020	0.015	-0.015	+0.030	0.015	-0.020	+0.050	0.050	-0.030	+0.070	0.050	-0.050	+0.070	0.110	-0.050
250	315	+0.023	0.020	-0.015	+0.030	0.020	-0.023	+0.050	0.060	-0.030	+0.080	0.060	-0.050	+0.080	0.130	-0.050
315	400	+0.025	0.025	-0.018	+0.036	0.025	-0.025	+0.060	0.080	-0.040	+0.090	0.080	-0.060	+0.090	0.150	-0.060

Nominal Size Range Millimetres		Class RC6 Medium Running			Class RC7 Free Running			Class RC8 Loose Running			Class RC9 Loose Running		
		Hole Tol. H9	Minimum Clearance	Shaft Tol. e8	Hole Tol. H9	Minimum Clearance	Shaft Tol. d8	Hole Tol. H10	Minimum Clearance	Shaft Tol. e9	Hole Tol. GR11	Minimum Clearance	Shaft Tol. gr10
Over	To	-0		+0	-0		+0	-0		+0	-0		+0
0	3	+0.025	0.015	-0.015	+0.025	0.025	-0.015	+0.041	0.064	-0.025	+0.060	0.100	-0.040
3	6	+0.030	0.015	-0.018	+0.030	0.030	-0.018	+0.046	0.071	-0.030	+0.080	0.110	-0.050
6	10	+0.036	0.025	-0.023	+0.036	0.040	-0.023	+0.056	0.076	-0.036	+0.070	0.130	-0.060
10	18	+0.040	0.030	-0.025	+0.040	0.050	-0.025	+0.070	0.090	-0.040	+0.100	0.150	-0.070
18	30	+0.050	0.040	-0.030	+0.050	0.060	-0.030	+0.090	0.110	-0.050	+0.130	0.180	-0.090
30	50	+0.060	0.050	-0.040	+0.060	0.080	-0.040	+0.100	0.130	-0.060	+0.150	0.200	-0.100
50	80	+0.080	0.060	-0.050	+0.080	0.100	-0.050	+0.110	0.150	-0.080	+0.180	0.230	-0.120
80	120	+0.090	0.080	-0.060	+0.090	0.130	-0.060	+0.130	0.180	-0.090	+0.230	0.250	-0.130
120	180	+0.100	0.090	-0.060	+0.100	0.150	-0.060	+0.150	0.200	-0.100	+0.250	0.300	-0.150
180	250	+0.110	0.100	-0.070	+0.110	0.180	-0.070	+0.180	0.250	-0.110	+0.300	0.380	-0.180
250	315	+0.130	0.130	-0.080	+0.130	0.200	-0.080	+0.200	0.300	-0.130	+0.300	0.460	-0.200
315	400	+0.150	0.150	-0.090	+0.150	0.250	-0.090	+0.230	0.360	-0.150	+0.360	0.560	-0.230

Running and sliding fits

VALUES IN THOUSANDTHS OF AN INCH

Nominal Size Range Inches		Class LC1			Class LC2			Class LC3			Class LC4			Class LC5			Class LC6		
		Hole Tol. GR6	Minimum Clearance	Shaft Tol. GR5	Hole Tol. GR8	Minimum Clearance	Shaft Tol. GR7	Hole Tol. GR10	Minimum Clearance	Shaft Tol. GR9	Hole Tol. GR7	Minimum Clearance	Shaft Tol. GR6	Hole Tol. GR9	Minimum Clearance	Shaft Tol. GR8	Hole Tol. GR9	Minimum Clearance	Shaft Tol. GR8
Over	To	-0		+0	-0		+0	-0		+0	-0		+0	-0		+0	-0		+0
0	.12	+0.25	0	-0.15	+0.4	0	-0.25	+0.6	0	-0.4	+1.6	0	-1.0	+0.4	0.10	-0.25	+1.0	0.3	-0.6
.12	.24	+0.30	0	-0.20	+0.5	0	-0.30	+0.7	0	-0.5	+1.8	0	-1.2	+0.5	0.15	-0.30	+1.2	0.4	-0.7
.24	.40	+0.40	0	-0.25	+0.6	0	-0.40	+0.9	0	-0.6	+2.2	0	-1.4	+0.6	0.20	-0.40	+1.4	0.5	-0.9
.40	.71	+0.40	0	-0.30	+0.7	0	-0.40	+1.0	0	-0.7	+2.8	0	-1.6	+0.7	0.25	-0.40	+1.6	0.6	-1.0
.71	1.19	+0.50	0	-0.40	+0.8	0	-0.50	+1.2	0	-0.8	+3.5	0	-2.0	+0.8	0.30	-0.50	+2.0	0.8	-1.2
1.19	1.97	+0.60	0	-0.40	+1.0	0	-0.60	+1.6	0	-1.0	+4.0	0	-2.5	+1.0	0.40	-0.60	+2.5	1.0	-1.6
1.97	3.15	+0.70	0	-0.50	+1.2	0	-0.70	+1.8	0	-1.2	+4.5	0	-3.0	+1.2	0.40	-0.70	+3.0	1.2	-1.8
3.15	4.73	+0.90	0	-0.60	+1.4	0	-0.90	+2.7	0	-1.4	+5.0	0	-3.5	+1.4	0.50	-0.90	+3.5	1.4	-2.2
4.73	7.09	+1.00	0	-0.70	+1.6	0	-1.00	+2.5	0	-1.6	+6.0	0	-4.0	+1.6	0.60	-1.00	+4.0	1.6	-2.5
7.09	9.85	+1.20	0	-0.80	+1.8	0	-1.20	+2.8	0	-1.8	+7.0	0	-4.5	+1.8	0.60	-1.20	+4.5	2.0	-2.8
9.85	12.41	+1.20	0	-0.90	+2.0	0	-1.20	+3.0	0	-2.0	+8.0	0	-5.0	+2.0	0.70	-1.20	+5.0	2.2	-3.0
12.41	15.75	+1.40	0	-1.00	+2.2	0	-1.40	+3.5	0	-2.2	+9.0	0	-6.0	+2.2	0.70	-1.40	+6.0	2.5	-3.5

Nominal Size Range Inches		Class LC7			Class LC8			Class LC9			Class LC10			Class LC11		
		Hole Tol. GR10	Minimum Clearance	Shaft Tol. GR9	Hole Tol. GR10	Minimum Clearance	Shaft Tol. GR9	Hole Tol. GR11	Minimum Clearance	Shaft Tol. GR10	Hole Tol. GR12	Minimum Clearance	Shaft Tol. GR11	Hole Tol. GR13	Minimum Clearance	Shaft Tol. GR12
Over	To	-0		+0	-0		+0	-0		+0	-0		+0	-0		+0
0	.12	+1.6	0.6	-1.0	+1.6	1.0	-1.0	+2.5	2.5	-1.6	+1.0	4.0	-2.5	+6.0	5.0	-4.0
.12	.24	+1.8	0.8	-1.2	+1.8	1.2	-1.2	+3.0	2.8	-1.8	+5.0	4.5	-3.0	+7.0	6.0	-5.0
.24	.40	+2.2	1.0	-1.4	+2.2	1.6	-1.4	+3.5	3.0	-2.2	+6.0	5.0	-3.5	+9.0	7.0	-6.0
.40	.71	+2.8	1.2	-1.6	+2.8	2.0	-1.6	+4.0	3.5	-2.8	+7.0	6.0	-4.0	+10.0	8.0	-7.0
.71	1.19	+3.5	1.6	-2.0	+3.5	2.5	-2.0	+5.0	4.5	-3.5	+8.0	7.0	-5.0	+12.0	10.0	-8.0
1.19	1.97	+4.0	2.0	-2.5	+4.0	3.6	-2.5	+6.0	5.0	-4.0	+10.0	8.0	-6.0	+16.0	12.0	-10.0
1.97	3.15	+4.5	2.5	-3.0	+4.5	4.0	-3.0	+7.0	6.0	-4.5	+12.0	10.0	-7.0	+18.0	14.0	-12.0
3.15	4.73	+5.0	3.0	-3.5	+5.0	5.0	-3.5	+9.0	7.0	-5.0	+14.0	11.0	-9.0	+22.0	16.0	-14.0
4.73	7.09	+6.0	3.5	-4.0	+6.0	6.0	-4.0	+10.0	8.0	-6.0	+16.0	12.0	-10.0	+25.0	18.0	-16.0
7.09	9.85	+7.0	4.0	-4.5	+7.0	7.0	-4.5	+12.0	10.0	-7.0	+18.0	16.0	-12.0	+28.0	22.0	-18.0
9.85	12.41	+8.0	4.5	-5.0	+8.0	7.0	-5.0	+12.0	12.0	-8.0	+20.0	20.0	-12.0	+30.0	28.0	-20.0
12.41	15.75	+9.0	5.0	-6.0	+9.0	8.0	-6.0	+14.0	14.0	-9.0	+22.0	22.0	-14.0	+35.0	30.0	-22.0

VALUES IN MILLIMETRES

Nominal Size Range Millimetres		Class LC1			Class LC2			Class LC3			Class LC4			Class LC5			Class LC6		
		Hole Tol. H6	Minimum Clearance	Shaft Tol. h5	Hole Tol. H7	Minimum Clearance	Shaft Tol. h6	Hole Tol. H8	Minimum Clearance	Shaft Tol. h7	Hole Tol. H10	Minimum Clearance	Shaft Tol. h9	Hole Tol. H7	Minimum Clearance	Shaft Tol. g6	Hole Tol. H9	Minimum Clearance	Shaft Tol. f8
Over	To	-0		+0	-0		+0	-0		+0	-0		+0	-0		+0	-0		+0
0	3	+0.006	0	-0.004	+0.010	0	-0.006	+0.015	0	-0.010	+0.041	0	-0.025	+0.010	0.002	-0.006	+0.025	0.008	-0.015
3	6	+0.008	0	-0.005	+0.013	0	-0.008	+0.018	0	-0.013	+0.046	0	-0.030	+0.013	0.004	-0.008	+0.030	0.010	-0.018
6	10	+0.010	0	-0.006	+0.015	0	-0.010	+0.023	0	-0.015	+0.056	0	-0.036	+0.015	0.005	-0.010	+0.036	0.013	-0.023
10	18	+0.010	0	-0.008	+0.018	0	-0.010	+0.025	0	-0.018	+0.070	0	-0.040	+0.018	0.006	-0.010	+0.041	0.015	-0.025
18	30	+0.013	0	-0.010	+0.020	0	-0.013	+0.030	0	-0.020	+0.090	0	-0.050	+0.020	0.008	-0.013	+0.050	0.020	-0.030
30	50	+0.015	0	-0.010	+0.025	0	-0.015	+0.041	0	-0.025	+0.100	0	-0.060	+0.025	0.010	-0.015	+0.060	0.030	-0.040
50	80	+0.018	0	-0.013	+0.030	0	-0.018	+0.046	0	-0.030	+0.110	0	-0.080	+0.030	0.010	-0.018	+0.080	0.030	-0.050
80	120	+0.023	0	-0.015	+0.036	0	-0.023	+0.056	0	-0.036	+0.130	0	-0.080	+0.036	0.013	-0.023	+0.090	0.040	-0.060
120	180	+0.025	0	-0.018	+0.041	0	-0.025	+0.064	0	-0.041	+0.150	0	-0.100	+0.041	0.015	-0.025	+0.100	0.040	-0.060
180	250	+0.030	0	-0.020	+0.046	0	-0.030	+0.071	0	-0.046	+0.180	0	-0.110	+0.046	0.015	-0.030	+0.110	0.050	-0.070
250	315	+0.020	0	-0.023	+0.051	0	-0.030	+0.076	0	-0.051	+0.200	0	-0.130	+0.051	0.018	-0.030	+0.130	0.060	-0.080
315	400	+0.036	0	-0.025	+0.056	0	-0.036	+0.089	0	-0.056	+0.230	0	-0.150	+0.056	0.018	-0.036	+0.150	0.060	-0.090

Nominal Size Range Millimetres		Class LC7			Class LC8			Class LC9			Class LC10			Class LC11		
		Hole Tol. H10	Minimum Clearance	Shaft Tol. e9	Hole Tol. H10	Minimum Clearance	Shaft Tol. d9	Hole Tol. H11	Minimum Clearance	Shaft Tol. c10	Hole Tol. GR12	Minimum Clearance	Shaft Tol. gr11	Hole Tol. GR13	Minimum Clearance	Shaft Tol. gr12
Over	To	-0		+0	-0		+0	-0		+0	-0		+0	-0		+0
0	3	+0.041	0.015	-0.025	+0.041	0.025	-0.025	+0.064	0.06	-0.041	+0.10	0.10	-0.06	+0.15	0.13	-0.10
3	6	+0.046	0.020	-0.030	+0.046	0.030	-0.030	+0.076	0.07	-0.046	+0.13	0.11	-0.08	+0.18	0.15	-0.13
6	10	+0.056	0.025	-0.036	+0.056	0.041	-0.036	+0.089	0.08	-0.056	+0.15	0.13	-0.09	+0.23	0.18	-0.15
10	18	+0.070	0.030	-0.040	+0.070	0.050	-0.040	+0.100	0.09	-0.070	+0.18	0.15	-0.10	+0.25	0.20	-0.18
18	30	+0.090	0.040	-0.050	+0.090	0.060	-0.050	+0.130	0.11	-0.090	+0.20	0.18	-0.13	+0.31	0.25	-0.20
30	50	+0.100	0.050	-0.060	+0.100	0.090	-0.060	+0.150	0.13	-0.100	+0.25	0.20	-0.15	+0.41	0.31	-0.25
50	80	+0.110	0.060	-0.080	+0.110	0.100	-0.080	+0.180	0.15	-0.110	+0.31	0.25	-0.18	+0.46	0.36	-0.31
80	120	+0.130	0.080	-0.090	+0.130	0.130	-0.090	+0.230	0.18	-0.130	+0.36	0.28	-0.23	+0.56	0.41	-0.36
120	180	+0.150	0.090	-0.100	+0.150	0.150	-0.100	+0.250	0.20	-0.150	+0.41	0.31	-0.25	+0.64	0.46	-0.41
180	250	+0.180	0.100	-0.110	+0.180	0.180	-0.110	+0.310	0.25	-0.180	+0.46	0.41	-0.31	+0.71	0.56	-0.46
250	315	+0.200	0.110	-0.130	+0.200	0.180	-0.130	+0.310	0.31	-0.200	+0.51	0.51	-0.31	+0.76	0.71	-0.51
315	400	+0.230	0.130	-0.150	+0.230	0.200	-0.150	+0.360	0.36	-0.230	+0.56	0.56	-0.36	+0.89	0.76	-0.56

Locational clearance fits

VALUES IN THOUSANDTHS OF AN INCH

Nominal Size Range Inches		Class LT1			Class LT2			Class LT3			Class LT4			Class LT5			Class LT6		
		Hole Tol. GR7	Maximum Interference	Shaft Tol. GR6	Hole Tol. GR8	Maximum Interference	Shaft Tol. GR7	Hole Tol. GR7	Maximum Interference	Shaft Tol. GR6	Hole Tol. GR8	Maximum Interference	Shaft Tol. GR7	Hole Tol. GR7	Maximum Interference	Shaft Tol. GR6	Hole Tol. GR8	Maximum Interference	Shaft Tol. GR7
Over	To	-0		+0	-0		+0	-0		+0	-0		+0	-0		+0	-0		+0
0	.12	+0.4	0.10	-0.25	+0.6	0.20	-0.4	+0.4	0.25	-0.25	+0.6	0.4	-0.4	+0.4	0.5	-0.25	+0.6	0.65	-0.4
.12	.24	+0.5	0.15	-0.30	+0.7	0.25	-0.5	+0.5	0.40	-0.30	+0.7	0.6	-0.5	+0.5	0.6	-0.30	+0.7	0.80	-0.5
.24	.40	+0.6	0.20	-0.40	+0.9	0.30	-0.6	+0.6	0.50	-0.40	+0.9	0.7	-0.6	+0.6	0.8	-0.40	+0.9	1.00	-0.6
.40	.71	+0.7	0.20	-0.40	+1.0	0.30	-0.7	+0.7	0.50	-0.40	+1.0	0.8	-0.7	+0.7	0.9	-0.40	+1.0	1.20	-0.7
.71	1.19	+0.8	0.25	-0.50	+1.2	0.40	-0.8	+0.8	0.60	-0.50	+1.2	0.9	-0.8	+0.8	1.1	-0.50	+1.2	1.40	-0.8
1.19	1.97	+1.0	0.30	-0.60	+1.6	0.50	-1.0	+1.0	0.70	-0.60	+1.6	1.1	-1.0	+1.0	1.3	-0.60	+1.6	1.70	-1.0
1.97	3.15	+1.2	0.30	-0.70	+1.8	0.60	-1.2	+1.2	0.80	-0.70	+1.8	1.3	-1.2	+1.2	1.5	-0.70	+1.8	2.00	-1.2
3.15	4.73	+1.4	0.40	-0.90	+2.2	0.70	-1.4	+1.4	1.00	-0.90	+2.2	1.5	-1.4	+1.4	1.9	-0.90	+2.2	2.40	-1.4
4.73	7.09	+1.6	0.50	-1.00	+2.5	0.80	-1.6	+1.6	1.10	-1.00	+2.5	1.7	-1.6	+1.6	2.2	-1.00	+2.5	2.80	-1.6
7.09	9.85	+1.8	0.60	-1.20	+2.8	0.90	-1.8	+1.8	1.40	-1.20	+2.8	2.0	-1.8	+1.8	2.6	-1.20	+2.8	3.20	-1.8
9.85	12.41	+2.0	0.60	-1.20	+3.0	1.00	-2.0	+2.0	1.40	-1.20	+3.0	2.2	-2.0	+2.0	2.6	-1.20	+3.0	3.40	-2.0
12.41	15.75	+2.2	0.70	-1.40	+3.5	1.00	-2.2	+2.2	1.60	-1.40	+3.5	2.4	-2.2	+2.2	3.0	-1.40	+3.5	3.80	-2.2

VALUES IN MILLIMETRES

Nominal Size Range Millimetres		Class LT1			Class LT2			Class LT3			Class LT4			Class LT5			Class LT6		
		Hole Tol. H7	Maximum Interference	Shaft Tol. js6	Hole Tol. H8	Maximum Interference	Shaft Tol. js7	Hole Tol. H7	Maximum Interference	Shaft Tol. k6	Hole Tol. H8	Maximum Interference	Shaft Tol. k7	Hole Tol. H7	Maximum Interference	Shaft Tol. n6	Hole Tol. H8	Maximum Interference	Shaft Tol. n7
Over	To	-0		+0	-0		+0	-0		+0	-0		+0	-0		+0	-0		+0
0	3	+0.010	0.002	-0.006	+0.015	0.005	-0.010	+0.010	0.006	-0.006	+0.015	0.010	-0.010	+0.010	0.013	-0.006	+0.015	0.016	-0.010
3	6	+0.013	0.004	-0.008	+0.018	0.006	-0.013	+0.013	0.010	-0.008	+0.018	0.015	-0.013	+0.013	0.015	-0.008	+0.018	0.020	-0.013
6	10	+0.015	0.005	-0.010	+0.023	0.008	-0.015	+0.015	0.013	-0.010	+0.023	0.018	-0.015	+0.015	0.020	-0.010	+0.023	0.025	-0.015
10	18	+0.018	0.005	-0.010	+0.025	0.010	-0.018	+0.018	0.013	-0.010	+0.025	0.020	-0.018	+0.018	0.023	-0.010	+0.025	0.030	-0.018
18	30	+0.020	0.006	-0.013	+0.030	0.010	-0.020	+0.020	0.015	-0.013	+0.030	0.023	-0.020	+0.020	0.028	-0.013	+0.030	0.036	-0.020
30	50	+0.025	0.008	-0.015	+0.041	0.013	-0.025	+0.025	0.018	-0.015	+0.041	0.028	-0.025	+0.025	0.033	-0.015	+0.041	0.044	-0.025
50	80	+0.030	0.008	-0.018	+0.046	0.015	-0.030	+0.030	0.020	-0.018	+0.046	0.033	-0.030	+0.030	0.038	-0.018	+0.046	0.051	-0.030
80	120	+0.036	0.010	-0.023	+0.056	0.018	-0.036	+0.036	0.025	-0.023	+0.056	0.038	-0.036	+0.036	0.048	-0.023	+0.056	0.062	-0.036
120	180	+0.041	0.013	-0.025	+0.064	0.020	-0.041	+0.041	0.028	-0.025	+0.064	0.044	-0.041	+0.041	0.056	-0.025	+0.064	0.071	-0.041
180	250	+0.046	0.015	-0.030	+0.071	0.023	-0.046	+0.046	0.036	-0.030	+0.071	0.051	-0.046	+0.046	0.066	-0.030	+0.071	0.081	-0.046
250	315	+0.051	0.015	-0.030	+0.076	0.025	-0.051	+0.051	0.036	-0.030	+0.076	0.056	-0.051	+0.051	0.066	-0.030	+0.076	0.086	-0.051
315	400	+0.056	0.018	-0.036	+0.089	0.025	-0.056	+0.056	0.041	-0.036	+0.089	0.062	-0.056	+0.056	0.076	-0.036	+0.089	0.096	-0.056

Location transition fits

VALUES IN THOUSANDTHS OF AN INCH

Nominal Size Range Inches		Class LN1 Light Press Fit			Class LN2 Medium Press Fit			Class LN3 Heavy Press Fit			Class LN4			Class LN5			Class LN6		
		Hole Tol. GR6	Maximum Interference	Shaft Tol. GR5	Hole Tol. GR7	Maximum Interference	Shaft Tol. GR6	Hole Tol. GR7	Maximum Interference	Shaft Tol. GR6	Hole Tol. GR8	Maximum Interference	Shaft Tol. GR7	Hole Tol. GR9	Maximum Interference	Shaft Tol. GR8	Hole Tol. GR10	Maximum Interference	Shaft Tol. GR9
Over	To	-0		+0	-0		+0	-0		+0	-0		+0	-0		+0	-0		+0
0	.12	+0.25	0.40	-0.15	+0.4	0.65	-0.25	+0.4	0.75	-0.25	+0.6	1.2	-0.4	+1.0	1.8	-0.6	+1.6	3.0	-1.0
.12	.24	+0.30	0.50	-0.20	+0.5	0.80	-0.30	+0.5	0.90	-0.30	+0.7	1.5	-0.5	+1.2	2.3	-0.7	+1.8	3.6	-1.2
.24	.40	+0.40	0.65	-0.25	+0.6	1.00	-0.40	+0.6	1.20	-0.40	+0.9	1.8	-0.6	+1.4	2.8	-0.9	+2.2	4.4	-1.4
.40	.71	+0.40	0.70	-0.30	+0.7	1.10	-0.40	+0.7	1.40	-0.40	+1.0	2.2	-0.7	+1.6	3.4	-1.0	+2.8	5.6	-1.6
.71	1.19	+0.50	0.90	-0.40	+0.8	1.30	-0.50	+0.8	1.70	-0.50	+1.2	2.6	-0.8	+2.0	4.2	-1.2	+3.5	7.0	-2.0
1.19	1.97	+0.60	1.00	-0.40	+1.0	1.60	-0.60	+1.0	2.00	-0.60	+1.6	3.4	-1.0	+2.5	5.3	-1.6	+4.0	8.5	-2.5
1.97	3.15	+0.70	1.30	-0.50	+1.2	2.10	-0.70	+1.2	2.30	-0.70	+1.8	4.0	-1.2	+3.0	6.3	-1.8	+4.5	10.0	-3.0
3.15	4.73	+0.90	1.60	-0.60	+1.4	2.50	-0.90	+1.4	2.90	-0.90	+2.2	4.8	-1.4	+4.0	7.7	-2.2	+5.0	11.5	-3.5
4.73	7.09	+1.00	1.90	-0.70	+1.6	2.80	-1.00	+1.6	3.50	-1.00	+2.5	5.6	-1.6	+4.5	8.7	-2.5	+6.0	13.5	-4.0
7.09	9.85	+1.20	2.20	-0.80	+1.8	3.20	-1.20	+1.8	4.20	-1.20	+2.8	6.6	-1.8	+5.0	10.3	-2.8	+7.0	16.5	-4.5
9.85	12.41	+1.20	2.30	-0.90	+2.0	3.40	-1.20	+2.0	4.70	-1.20	+3.0	7.5	-2.0	+6.0	12.0	-3.0	+8.0	19.0	-5.0
12.41	15.75	+1.40	2.60	-1.00	+2.2	3.90	-1.40	+2.2	5.90	-1.40	+3.5	8.7	-2.2	+6.0	14.5	-3.5	+9.0	23.0	-6.0

VALUES IN MILLIMETRES

Nominal Size Range Millimetres		Class LN1 Light Press Fit			Class LN2 Medium Press Fit			Class LN3 Heavy Press Fit			Class LN4			Class LN5			Class LN6		
		Hole Tol. GR6	Maximum Interference	Shaft Tol. gr5	Hole Tol. H7	Maximum Interference	Shaft Tol. p6	Hole Tol. H7	Maximum Interference	Shaft Tol. t6	Hole Tol. GR8	Maximum Interference	Shaft Tol. gr7	Hole Tol. GR9	Maximum Interference	Shaft Tol. gr7	Hole Tol. GR10	Maximum Interference	Shaft Tol. gr9
Over	To	-0		+0	-0		+0	-0		+0	-0		+0	-0		+0	-0		+0
0	3	+0.006		-0.004	+0.010	0.016	-0.006	+0.010	0.019	-0.006	+0.015	0.030	-0.010	+0.025	0.046	-0.015	+0.041	0.076	-0.025
3	6	+0.008		-0.005	+0.013	0.020	-0.008	+0.013	0.023	-0.008	+0.018	0.038	-0.013	+0.030	0.059	-0.018	+0.046	0.091	-0.030
6	10	+0.010		-0.006	+0.015	0.025	-0.010	+0.015	0.030	-0.010	+0.023	0.046	-0.015	+0.036	0.071	-0.023	+0.056	0.112	-0.036
10	18	+0.010		-0.008	+0.018	0.028	-0.010	+0.018	0.036	-0.010	+0.025	0.056	-0.018	+0.041	0.086	-0.025	+0.071	0.142	-0.041
18	30	+0.013		-0.010	+0.020	0.033	-0.013	+0.020	0.044	-0.013	+0.030	0.066	-0.020	+0.051	0.107	-0.030	+0.089	0.178	-0.051
30	50	+0.015		-0.010	+0.025	0.041	-0.015	+0.025	0.051	-0.015	+0.041	0.086	-0.025	+0.064	0.135	-0.041	+0.102	0.216	-0.064
50	80	+0.018		-0.013	+0.030	0.054	-0.018	+0.030	0.059	-0.018	+0.046	0.102	-0.030	+0.076	0.160	-0.046	+0.114	0.254	-0.076
80	120	+0.023		-0.015	+0.036	0.064	-0.023	+0.036	0.074	-0.023	+0.056	0.122	-0.036	+0.102	0.196	-0.056	+0.127	0.292	-0.102
120	180	+0.025		-0.018	+0.041	0.071	-0.025	+0.041	0.089	-0.025	+0.064	0.142	-0.041	+0.114	0.221	-0.064	+0.152	0.343	-0.114
180	250	+0.030		-0.020	+0.046	0.081	-0.030	+0.046	0.107	-0.030	+0.071	0.168	-0.046	+0.127	0.262	-0.071	+0.178	0.419	-0.127
250	315	+0.030		-0.023	+0.051	0.086	-0.030	+0.051	0.119	-0.030	+0.076	0.191	-0.051	+0.152	0.305	-0.076	+0.203	0.483	-0.152
315	400	+0.036		-0.025	+0.056	0.099	-0.036	+0.056	0.150	-0.036	+0.089	0.221	-0.056	+0.152	0.368	-0.089	+0.229	0.584	-0.152

Locational interference fits

VALUES IN THOUSANDTHS OF AN INCH

Nominal Size Range Inches		Class FN1 Light Drive Fit			Class FN2 Medium Drive Fit			Class FN3 Heavy Drive Fit			Class FN4 Shrink Fit			Class FN5 Heavy Shrink Fit		
		Hole Tol. GR6	Maximum Interference	Shaft Tol. GR5	Hole Tol. GR7	Maximum Interference	Shaft Tol. GR6	Hole Tol. GR7	Maximum Interference	Shaft Tol. GR6	Hole Tol. GR7	Maximum Interference	Shaft Tol. GR6	Hole Tol. GR8	Maximum Interference	Shaft Tol. GR7
Over	To	-0		+0	-0		+0	-0		+0	-0		+0	-0		+0
0	.12	+0.25	0.50	-0.15	+0.40	0.85	-0.25				+0.40	0.95	-0.25	+0.60	1.30	-0.40
.12	.24	+0.30	0.60	-0.20	+0.50	1.00	-0.30				+0.50	1.20	-0.30	+0.70	1.70	-0.50
.24	.40	+0.40	0.75	-0.25	+0.60	1.40	-0.40				+0.60	1.60	-0.40	+0.90	2.00	-0.60
.40	.56	+0.40	0.80	-0.30	+0.70	1.60	-0.40				+0.70	1.80	-0.40	+1.00	2.30	-0.70
.56	.71	+0.40	0.90	-0.30	+0.70	1.60	-0.40				+0.70	1.80	-0.40	+1.00	2.50	-0.70
.71	.95	+0.50	1.10	-0.40	+0.80	1.90	-0.50				+0.80	2.10	-0.50	+1.20	3.00	-0.80
.95	1.19	+0.50	1.20	-0.40	+0.80	1.90	-0.50	+0.80	2.10	-0.50	+0.80	2.30	-0.50	+1.20	3.30	-0.80
1.19	1.58	+0.60	1.30	-0.40	+1.00	2.40	-0.60	+1.00	2.60	-0.60	+1.00	3.10	-0.60	+1.60	4.00	-1.00
1.58	1.97	+0.60	1.40	-0.40	+1.00	2.40	-0.60	+1.00	2.80	-0.60	+1.00	3.40	-0.60	+1.60	5.00	-1.00
1.97	2.56	+0.70	1.80	-0.50	+1.20	2.70	-0.70	+1.20	3.20	-0.70	+1.20	4.20	-0.70	+1.80	6.20	-1.20
2.56	3.15	+0.70	1.90	-0.50	+1.20	2.90	-0.70	+1.20	3.70	-0.70	+1.20	4.70	-0.70	+1.80	7.20	-1.20
3.15	3.94	+0.90	2.40	-0.60	+1.40	3.70	-0.90	+1.40	4.40	-0.70	+1.40	5.90	-0.90	+2.20	8.40	-1.40

VALUES IN MILLIMETRES

Nominal Size Range Millimetres		Class FN1 Light Drive Fit			Class FN2 Medium Drive Fit			Class FN3 Heavy Drive Fit			Class FN4 Shrink Fit			Class FN5 Heavy Shrink Fit		
		Hole Tol. GR6	Maximum Interference	Shaft Tol. gr5	Hole Tol. H7	Maximum Interference	Shaft Tol. s6	Hole Tol. H7	Maximum Interference	Shaft Tol. t6	Hole Tol. GR8	Maximum Interference	Shaft Tol. gr7	Hole Tol. H8	Maximum Interference	Shaft Tol. t7
Over	To	-0		+0	-0		+0	-0		+0	-0		+0	-0		+0
0	3	+0.006	0.013	-0.004	+0.010	0.216	-0.006				+0.010	0.024	-0.006	+0.015	0.033	-0.010
3	6	+0.007	0.015	-0.005	+0.013	0.025	-0.007				+0.013	0.030	-0.007	+0.018	0.043	-0.013
6	10	+0.010	0.019	-0.006	+0.015	0.036	-0.010				+0.015	0.041	-0.010	+0.023	0.051	-0.015
10	14	+0.010	0.020	-0.008	+0.018	0.041	-0.010				+0.018	0.046	-0.010	+0.025	0.058	-0.018
14	18	+0.010	0.023	-0.008	+0.018	0.041	-0.010				+0.018	0.046	-0.010	+0.025	0.064	-0.018
18	24	+0.013	0.028	-0.010	+0.020	0.048	-0.013				+0.020	0.053	-0.013	+0.030	0.076	-0.020
24	30	+0.013	0.030	-0.010	+0.020	0.048	-0.013	+0.020	0.053	-0.013	+0.020	0.058	-0.013	+0.030	0.084	-0.020
30	40	+0.015	0.033	-0.010	+0.025	0.061	-0.015	+0.025	0.066	-0.015	+0.025	0.079	-0.015	+0.041	0.102	-0.025
40	50	+0.015	0.036	-0.010	+0.025	0.061	-0.015	+0.025	0.071	-0.015	+0.025	0.086	-0.015	+0.041	0.127	-0.025
50	65	+0.018	0.046	-0.013	+0.030	0.069	-0.018	+0.030	0.082	-0.018	+0.030	0.107	-0.018	+0.046	0.157	-0.030
65	80	+0.018	0.048	-0.013	+0.030	0.074	-0.018	+0.030	0.094	-0.018	+0.030	0.119	-0.018	+0.046	0.183	-0.030
80	100	+0.023	0.061	-0.015	+0.035	0.094	-0.023	+0.035	0.112	-0.023	+0.036	0.150	-0.023	+0.056	0.213	-0.036

Force and shrink fits

APPENDIX D

DRILLED HOLE TOLERANCE (UNDER NORMAL SHOP CONDITIONS)

STANDARD DRILL SIZE				TOLERANCE IN DECIMALS	
DRILL SIZE				PLUS	MINUS
Number	Fraction	Decimal	Metric (MM)		
80		0.0135	0.3412	0.0023	
79		0.0145	0.3788	0.0024	
—	1/64	0.0156	0.3969	0.0025	
78		0.0160	0.4064	0.0025	
77		0.0180	0.4572	0.0026	
76		0.0200	0.5080	0.0027	
75		0.0210	0.5334	0.0027	
74		0.0225	0.5631	0.0028	
73		0.0240	0.6096	0.0028	
72		0.0250	0.6350	0.0029	
71		0.0260	0.6604	0.0029	
70		0.0280	0.7112	0.0030	.0005
69		0.0292	0.7483	0.0030	
68		0.0310	0.7874	0.0031	
—	1/32	0.0312	0.7937	0.0031	
67		0.0320	0.8128	0.0031	
66		0.0330	0.8382	0.0032	
65		0.0350	0.8890	0.0032	
64		0.0360	0.9144	0.0033	
63		0.0370	0.9398	0.0033	
62		0.0380	0.9652	0.0033	
61		0.0390	0.9906	0.0033	
60		0.0400	1.0160	0.0034	
59		0.0410	1.0414	0.0034	
58		0.0420	1.0668	0.0034	
57		0.0430	1.0922	0.0035	
56		0.0465	1.1684	0.0035	
—	3/64	0.0469	1.1906	0.0036	
55		0.0520	1.3208	0.0037	
54		0.0550	1.3970	0.0038	
53		0.0595	1.5122	0.0039	
—	1/16	0.0625	1.5875	0.0039	
52		0.0635	1.6002	0.0039	
51		0.0670	1.7018	0.0040	
50		0.0700	1.7780	0.0041	
49		0.0730	1.8542	0.0041	.001
48		0.0760	1.9304	0.0042	
—	5/64	0.0781	1.9844	0.0042	
47		0.0785	2.0001	0.0042	
46		0.0810	2.0574	0.0043	
45		0.0820	2.0828	0.0043	
44		0.0860	2.1844	0.0044	
43		0.0890	2.2606	0.0044	
42		0.0935	2.3622	0.0045	
—	3/32	0.0937	2.3812	0.0045	
41		0.0960	2.4384	0.0045	
40		0.0980	2.4892	0.0046	
39		0.0995	2.5377	0.0046	
38		0.1015	2.5908	0.0046	
37		0.1040	2.6416	0.0047	
36		0.1065	2.6924	0.0047	
—	7/64	0.1094	2.7781	0.0047	

DRILLED HOLE TOLERANCE (UNDER NORMAL SHOP CONDITIONS)

STANDARD DRILL SIZE				TOLERANCE IN DECIMALS	
DRILL SIZE				PLUS	MINUS
No./Letter	Fraction	Decimal	Metric (MM)		
35		0.1100	2.7490	0.0047	
34		0.1110	2.8194	0.0048	
33		0.1130	2.8702	0.0048	
32		0.1160	2.9464	0.0048	
31		0.1200	3.0480	0.0049	
—	1/8	0.1250	3.1750	0.0050	
30		0.1285	3.2766	0.0050	
29		0.1360	3.4544	0.0051	
28		0.1405	3.5560	0.0052	
—	9/64	0.1406	3.5719	0.0052	
27		0.1440	3.6576	0.0052	
26		0.1470	3.7338	0.0052	
25		0.1495	3.7886	0.0053	
24		0.1520	3.8608	0.0053	
23		0.1540	3.9116	0.0053	
—	5/32	0.1562	3.9687	0.0053	
22		0.1570	3.9878	0.0053	
21		0.1590	4.0386	0.0054	
20		0.1610	4.0894	0.0054	
19		0.1660	4.2164	0.0055	
18		0.1695	4.3180	0.0055	
—	11/64	0.1719	4.3656	0.0055	
17		0.1730	4.3942	0.0055	
16		0.1770	4.4958	0.0056	.001
15		0.1800	4.5720	0.0056	
14		0.1820	4.6228	0.0057	
13		0.1850	4.6990	0.0057	
—	3/16	0.1875	4.7625	0.0057	
12		0.1890	4.8006	0.0057	
11		0.1910	4.8514	0.0057	
10		0.1935	4.9276	0.0058	
9		0.1960	4.9784	0.0058	
8		0.1990	5.0800	0.0058	
7		0.2010	5.1054	0.0058	
—	13/64	0.2031	5.1594	0.0058	
6		0.2040	5.1816	0.0058	
5		0.2055	5.2070	0.0059	
4		0.2090	5.3086	0.0059	
3		0.2130	5.4102	0.0059	
—	7/32	0.2187	5.5562	0.0060	
2		0.2210	5.6134	0.0060	
1		0.2280	5.7912	0.0061	
A		0.2340	5.9436	0.0061	
—	15/64	0.2344	5.9531	0.0061	
B		0.2380	6.0452	0.0061	
C		0.2420	6.1468	0.0062	
D		0.2460	6.2484	0.0062	
E	1/4	0.2500	6.3500	0.0063	
F		0.2570	6.5278	0.0063	
G		0.2610	6.6294	0.0063	
—	17/64	0.2656	6.7469	0.0064	
H		0.2660	6.7564	0.0064	
I		0.2720	6.9088	0.0064	.002
J		0.2770	7.0358	0.0065	
K		0.2810	7.1374	0.0065	
—	9/32	0.2812	7.1437	0.0065	
L		0.2900	7.3660	0.0066	
M		0.2950	7.4930	0.0066	
—	19/64	0.2969	7.5406	0.0066	

DRILLED HOLE TOLERANCE (UNDER NORMAL SHOP CONDITIONS)

STANDARD DRILL SIZE				TOLERANCE IN DECIMALS	
DRILL SIZE				PLUS	MINUS
Letter	Fraction	Decimal	Metric (MM)		
N		0.3020	7.6708	0.0067	
—	5/16	0.3125	7.9375	0.0067	
O		0.3160	8.0264	0.0068	
P		0.3230	8.2042	0.0068	
—	21/64	0.3281	8.3344	0.0068	
Q		0.3320	8.4328	0.0069	
R		0.3390	8.6106	0.0069	
—	11/32	0.3437	8.7312	0.0070	
S		0.3480	8.8392	0.0070	
T		0.3580	9.0932	0.0071	
—	23/64	0.3594	9.1281	0.0071	
U		0.3680	9.3472	0.0071	
—	3/8	0.3750	9.5250	0.0072	
V		0.3770	9.5758	0.0072	
W		0.3860	9.8044	0.0072	
—	25/64	0.3906	9.9219	0.0073	
X		0.3970	10.0838	0.0073	
Y		0.4040	10.2616	0.0073	
—	13/32	0.4062	10.3187	0.0074	
Z		0.4130	10.4902	0.0074	.002
	27/64	0.4219	10.7156	0.0075	
	7/16	0.4375	10.1125	0.0075	
	29/64	0.4531	11.5094	0.0076	
	15/32	0.4687	11.9062	0.0077	
	31/64	0.4844	12.3031	0.0078	
	1/2	0.5000	12.7000	0.0079	
	33/64	0.5156	13.0968	0.0080	
	17/32	0.5312	13.4937	0.0081	
	35/64	0.5469	13.8906	0.0081	
	9/16	0.5625	14.2875	0.0082	
	37/64	0.5781	14.6844	0.0083	
	19/32	0.5927	15.0812	0.0084	
	39/64	0.6094	15.4781	0.0084	
	5/8	0.6250	15.8750	0.0085	
	41/64	0.6406	16.2719	0.0086	
	21/32	0.6562	16.6687	0.0086	
	43/64	0.6719	17.0656	0.0087	
	11/16	0.6875	17.4625	0.0088	
	45/64	0.7031	17.8594	0.0088	
	23/32	0.7187	18.2562	0.0089	
	47/64	0.7344	18.6532	0.0090	
	3/4	0.7500	19.0500	0.0090	
	49/64	0.7656	19.4469	0.0091	
	25/32	0.7812	19.8433	0.0092	
	51/64	0.7969	20.2402	0.0092	
	13/16	0.8125	20.6375	0.0093	
	53/64	0.8281	21.0344	0.0093	
	27/32	0.8437	21.4312	0.0094	
	55/64	0.8594	21.8281	0.0095	
	7/8	0.8750	22.2250	0.0095	.003
	57/64	0.8906	22.6219	0.0096	
	29/32	0.9062	23.0187	0.0096	
	59/64	0.9219	23.4156	0.0097	
	15/16	0.9375	23.8125	0.0097	
	61/64	0.9531	24.2094	0.0098	
	31/32	0.9687	24.6062	0.0098	
	63/64	0.9844	25.0031	0.0099	
	1	1.0000	25.4000	0.0100	